Advanced Applications of Computational Mathematics

RIVER PUBLISHERS SERIES IN MATHEMATICAL AND ENGINEERING SCIENCES

Mathematics is the basis of all disciplines in science and engineering. Especially applied mathematics has become complementary to every branch of engineering sciences. The purpose of this book series is to present novel results in emerging research topics on engineering sciences, as well as to summarize existing research. It engrosses mathematicians, statisticians, scientists and engineers in a comprehensive range of research fields with different objectives and skills, such as differential equations, finite element method, algorithms, discrete mathematics, numerical simulation, machine leaning, probability and statistics, fuzzy theory, etc.

Books published in the series include professional research monographs, edited volumes, conference proceedings, handbooks and textbooks, which provide new insights for researchers, specialists in industry, and graduate students.

Topics covered in the series include, but are not limited to:

- Advanced mechatronics and robotics
- Artificial intelligence
- Automotive systems
- Discrete mathematics and computation
- Fault diagnosis and fault tolerance
- Finite element methods
- Fuzzy and possibility theory
- Industrial automation, process control and networked control systems
- Intelligent control systems
- Neural computing and machine learning
- Operations research and management science
- Optimization and algorithms
- Queueing systems
- Reliability, maintenance and safety for complex systems
- Resilience
- Stochastic modelling and statistical inference
- Supply chain management
- System engineering, control and monitoring
- Tele robotics, human computer interaction, human-robot interaction

For a list of other books in this series, visit www.riverpublishers.com

Advanced Applications of Computational Mathematics

Editors

Akshay Kumar
Graphic Era Hill University, India

Mangey Ram
Graphic Era Deemed to be University, India

Hari Mohan Srivastava
University of Victoria, Canada

Published 2021 by River Publishers
River Publishers
Alsbjergvej 10, 9260 Gistrup, Denmark
www.riverpublishers.com

Distributed exclusively by Routledge
4 Park Square, Milton Park, Abingdon, Oxon OX14 4RN
605 Third Avenue, New York, NY 10017, USA

Advanced Applications of Computational Mathematics / by Akshay Kumar, Mangey Ram, Hari Mohan Srivastava.

Routledge is an imprint of the Taylor & Francis Group, an informa business

ISBN 978-87-7022-605-9 (print)

Contents

Preface

In recent years, engineering and mathematical scientists worldwide have made significant and noteworthy advances on computational mathematics and its widespread applications. In fact, computational mathematics has become a focal point of study of modern-day engineering applications. The purpose of this book is to deliberate multidisciplinary studies involving original and advanced research in the field of computational and applied mathematics dealing with research methodology, techniques, applications, and algorithms. In the context of real-life problems, computational mathematics plays a key role and solves various problems in engineering and mathematical sciences. In this book, the following topics have been discussed:

- Locating ROI in iris using randomized Hough transform
- Subclasses of analytic functions defined by q-analogue of differential operator
- Scalar potential in planar Filippov systems
- Cross diffusion, nonlinear radiation, and Newtonian heating effects of third-grade fluid flow over a Riga plate with entropy generation minimization
- Canonical representation of Koga's solution to the Dirac equation
- Bounds of generalized "useful"information measure with application of Jensen's inequality
- Numerical methods and some of its application
- Bifurcation analysis for COVID-19 model with inhibitory effect
- Variable selection in bioinformatics: methods and algorithms
- Crime prediction via fuzzy multi-criteria decision-making approach under hesitant fuzzy environment
- Computational study of double diffusive MHDbuoyancy induced free convection in porous media with chemical reaction and internal heating
- Second law analysis of Williamson nanofluid flow in attendance of radiation and heat generation with Cattaneo-Christov (C-C) heat flux
- Importance of analytic continuation in complex analysis

List of Figures

List of Tables

List of Contributors

Arora, Geeta, *Lovely Professional University, Phagwara, India; E-mail: geetadma@gmail.com*

Bhushan, Mahak, *Indian Institute Science Education and Research, Kolkata, India; E-mail: mahakbhushan22@gmail.com*

Dutta, Palash, *Department of Mathematics, Dibrugarh University, Dibrugarh, India; E-mail: palash.dtt@gmail.com*

Dwivedi, Pankaj Prasad, *Jaypee University of Engineering and Technology, A.B. Road, Raghogarh, Guna, Madhya Pradesh, India*

Goala, Soumendra, *Department of Mathematics, Dibrugarh University, Dibrugarh, India; E-mail: soumendragoala@gmail.com*

Gulati, Paras, *Department of Computer Science & Engineering, Graphic Era Hill University, Dehradun, India; E-mail: parasgulati150@gmail.com*

Hanus, Tomáš, *University of Chemistry and Technology, Prague, Czech Republic; E-mail: tuark@seznam.cz*

Janovská, Drahoslava, *University of Chemistry and Technology, Prague, Czech Republic; E-mail: janovskd@vscht.cz*

Jayswal, Ekta, *Department of Mathematics, Gujarat University, Ahmedabad, Gujarat, India; E-mail: jayswal.ekta1993@gmail.com*

Kapoor, S., *Regional Institute of Education (NCERT), Bhubaneswar, Odisha, India; E-mail: saurabh09.ncert@gmail.com*

Kaur, J., *Department of Mathematic, Akal University, Talwandi Sabo, Bathinda, Punjab, India*

Kulkarni, Shubham, *Centre for Modeling and Simulation, SPPU, Pune, India; E-mail: kulkarnishubham01@gmail.com*

Limboo, Bulendra, *Department of Mathematics, Dibrugarh University, Dibrugarh, India; E-mail: rs_bulendralimboo@dibru.ac.in*

Loganathan, K., *Research and Development Wing, Live4Research, Tiruppur, Tamil Nadu, India; E-mail: loganathankaruppusamy304@gmail.com*

Modak, Sonal, *Lead Domain Engineer, Life Sciences and Healthcare Unit, Persistent Systems Inc., Santa Clara, CA, USA; E-mail: samurai.modak@gmail.com*

Mohanraj, M., *Department of Mechanical Engineering, Hindusthan College of Engineering and Technology, Coimbatore, Tamil Nadu, India*

Munjal, A., *Department of Mathematic, Akal University, Talwandi Sabo, Bathinda, Punjab, India*

Pandey, S. K., *Department of Mathematics, SPUP, Jodhpur (Raj.), India; E-mail: skpandey12@gmail.com*

Reddy, P. Thirupathi, *Department of Mathematics, Kakatiya University, Warangal, Telangana, India; E-mail: reddypt2@gmail.com*

Sagayaraj, A. Charles, *Department of Mathematics, Sri Vidya Mandir Arts and Science College, Katteri, Uthangarai, Tamil Nadu, India*

Selvamani, C., *Department of Mathematics, Faculty of Engineering, Karpagam Academy of Higher Education, Coimbatore, Tamil Nadu, India*

Shah, Nita H., *Department of Mathematics, Gujarat University, Ahmedabad, Gujarat, India; E-mail: nitahshah@gmail.com*

Sharma, D. K., *Jaypee University of Engineering and Technology, A.B. Road, Raghogarh, Guna, Madhya Pradesh, India; E-mail: dilipsharmajiet@gmail.com*

Sheoran, Nisha, *Department of Mathematics, Gujarat University, Ahmedabad, Gujarat, India; E-mail: sheorannisha@gmail.com*

Singh, Kamna, *Lovely Professional University, Phagwara, India; Email: kamnakamnasingh13@gmail.com,*

Singh, Teekam, *Department of Mathematics, Graphic Era Hill University, Dehradun, India; E-mail; tsingh@gehu.ac.in; tsingh@ma.iitr.ac.in*

Sridevi, S., *Department of Mathematics, GSS, GITAM University, Doddaballapur, Bengaluru Rural, Karnataka, India; E-mail: siri_settipalli@yahoo.co.in*

Sujatha, *Department of Mathematics, GSS, GITAM University, Doddaballapur, Bengaluru Rural, Karnataka, India; E-mail: sujathavaishnavy@gmail.com*

Tamilvanan, K., *Department of Mathematics, Government Arts College for Men, Krishnagiri, Tamil Nadu, India*

Valadi, Jayaraman, *Center for Informatics, School of Natural Sciences (SoNS), Shiv Nadar University, Greater Noida, Uttar Pradesh, India; and Computer Science Department, Flame University, Pune, India; E-mail: jayaraman.vk@flame.edu.in*

Venkateswarlu, B., *Department of Mathematics, GSS, GITAM University, Doddaballapur, Bengaluru Rural, Karnataka, India; E-mail: bvlmaths@gmail.comW9355*

List of Abbreviations

a	Stretching rate $\left(\mathrm{s}^{-1}\right)$
Be	Bejan number
Br	Brinkman number
C	Concentration $\left(\mathrm{kgm}^{-3}\right)$
C_r	Chemical reaction parameter
C_p	Specific heat $\left(\mathrm{J\ kg}^{-1}\mathrm{K}^{-1}\right)$
C_∞	Ambient concentration $\left(\mathrm{kgm}^{-3}\right)$
C_w	Fluid wall concentration $\left(\mathrm{kgm}^{-3}\right)$
Cf_x	Skin friction coefficient
D_F	Dufour number
$d(A, B)$	Distance measure
E_G	Entropy generation parameter
$F_1(\zeta)$	Velocity similarity function
H_g	Heat generation parameter
h_f	Convective heat transfer coefficient $\left(\mathrm{W\ m}^{-1}\mathrm{K}^{-1}\right)$
$h_A(x)$	Hesitant fuzzy element
l_h	Number of elements in h
k	Thermal conductivity $\left(\mathrm{W\ m}^{-1}\mathrm{K}^{-1}\right)$
Nu_x	Nusselt number
Nw	Newtonian heating parameter
Pr	Prandtl number
Q_0	Dimensional heat generation/absorption coefficient
Q	Modified Hartmann number
Rd	Radiation parameter
Re	Reynolds number
S_C	Schmidt number
$S(C)$	Satisfaction degree of C
$s(h)$	Score function of h
S_r	Soret number
Sh_x	Sherwood number
S_{gen}'''	Local volumetric entropy generation rate $\left(\mathrm{Wm}^{-3}\mathrm{K}^{-1}\right)$

$S_0^{'''}$	Characteristic entropy generation rate $\left(\mathrm{Wm^{-3}K^{-1}}\right)$
T	Temperature (K)
T_∞	Ambient temperature (K)
$\mathrm{u_w}$	Velocity of the sheet $\left(\mathrm{m\,s^{-1}}\right)$
u, v	Velocity components in (x, y) directions $\left(\mathrm{m\,s^{-1}}\right)$
x, y	Cartesian coordinates (m)
$\alpha_1,\ \alpha_2,\ \beta_1$	Fluid parameters
$\phi_1\,(\eta)$	Concentration similarity function
γ	Dimensionless thermal relaxation time
ζ	Similarity parameter
λ_T	Thermal relaxation time
λ	Dimensionless constant
ν	Kinematic viscosity $\left(\mathrm{m^2s^{-1}}\right)$
Ω	Dimensionless temperature difference
$\theta_1\,(\eta)$	Temperature similarity function
ρ	Density $\left(\mathrm{kgm^{-1}}\right)$
σ	Electrical conductivity $(\mathrm{S\,m})$
ψ	Stream function $\left(\mathrm{m\,s^{-1}}\right)$
ϕ	Dimensionless concentration difference
$\mu_A(x)$	Membership value/degree of x
$\overline{\sigma}(x)$	Deviation degree

ACO	Ant colony optimization
ANOVA	Analysis of variance
AUC	Area under curve
BBO	Biogeography-based optimization
BCI	Brain computer interface
BGUIMAJI	Bounds of generalized 'Useful' information measure with application of Jensen's inequality
CAST	cluster affinity search technique
CoEPrA	Comparative evaluation of prediction algorithms
DFE	Disease-free equilibrium
DLBCL	Diffuse large B-cell lymphomas
DLDA	Diagonal linear discriminant analysis
DLSR	Discriminative least square regression
EDA	Estimation of distribution algorithm
EE	Endemic equilibrium
et al.	And others
etc.	And more

FIFS	Frequent item feature selection
FST	fixation index
GA	Genetic algorithm
HFE	Hesitant fuzzy element
HFS	Hesitant fuzzy set
i.e.	That is
IFS	Intuitionistic fuzzy set
KNN	*k*-nearest neighbors
k-TSP	k-top scoring pair
LARS	Least angle regression
LOOCV	Leave one out cross-validation
LR	Logistic regression
MADM	Multi attribute decision-making
MAE	Mean absolute error
MCC	Matthew correlation coefficient
MCDM	Multi-criteria decision-making
MDD	Major depressive disorder
MHC-I	Major histocompatibility
MLL	Mixed lineage leukemia
MLP	Multilayer perceptron
mRMR	Minimum redundancy, maximum-relevancy
NCA	Neighborhood component analysis
NNC	Nearest neighbor classifier
NNE	Neural network ensemble
NSVA	Non-negative singular value approximation
ODE	ordinary differential equations
OOB	Out of bag
PAM	Partitioning around medoids
PBMC	Peripheral blood mononuclear cells
PSG	Polysomnographs
PSO	Particle swarm optimization
QSAR	Quantitative structure activity relationship
RBF	Radial basis function
RF	Random forest
ROC	Receiver operating characteristics
SEIQHR	Susceptible, Exposed, Infective, Quarantine, Hospitalized and Recover
SEIR	Susceptible, Exposed, Infective and Recover
SFFS	sequential forward floating search

SHHS	Sleep heart health study
SIRS	Susceptible, Infective, Recover and Susceptible
SNP	Single nucleiotide polymorphism
SPCA	Sparse principal component analysis
SPECTF	Single-Photon emission computed tomography features
SUQC	Susceptible, Un-quarantined, Quarantined and Confirmed
SVD	Singular value decomposition
SVM	Support vector machines
SVM-RFE	SVM recursive feature elimination
TPR	True positive rate

List of Symbols

ODE	Ordinary differential equation
X	Domain
S	State space
Σ_{12}	Boundary
$\triangle S_1, \triangle S_2, \triangle\Sigma_{12}$	Connected components of sets
$\mathbf{F}_1, \mathbf{F}_2$	Vector fields
$\overline{S}_1, \overline{S}_2$	Closures of sets
$\mathbf{\Sigma}^N(\mathbf{x})$	Normal vector
$\mathbf{\Sigma}^T(\mathbf{x})$	Tangent vector
$\mathbf{G}_{12}^{AM}$	Vector by arithmetic mean
$\mathbf{G}_{12}^{FM}$	Vector by Filippov method
Σ_{12}^{PD}	Points with parallel difference
Σ_{12}^{DT}	Points with double tangency
Σ_{12}^{TP}	Tangent points
$\kappa(\mathbf{x}),\ \tau(\mathbf{x}), \sigma(\mathbf{x}), \chi(\mathbf{x})$	Scalar functions on boundary
$\mathbf{\Phi}$	Parameterization of curve
$\mathcal{K}$	Curve
$\mathcal{L}$	Curve

$\mathcal{P}$	Curve
$U(\mathbf{x})$	Scalar potential
$\mathrm{proj}_{\mathbf{\Sigma}^T}\mathbf{F}_1(\mathbf{x})$	Orthogonal projection
$\mathrm{proj}_{\mathbf{\Sigma}^T}\mathbf{F}_2(\mathbf{x})$	Orthogonal projection
$G = (V, E)$	Graph with vertices and edges
$\mathrm{d}\mathbf{r} = [dx,\ dy]$	Differential of vector
$\mathcal{O}_\epsilon(x_0, y_0)$	Neighbourhood of point
$\mathbb{O}_\epsilon^R(x_0 \cdot y_0)$	Reduced neighbourhood of point
CE, CP, ONP, OFP, BE, TP	Common equilibrium, Crossing point, Onset point, Offset point, Boundary equilibrium, Tangent point
Σ_{12}^S	Sliding boundary
Σ_{12}^C	Crossing boundary
$\rho_1,\ \rho_2$	Planes
$\curvearrowleft,\ \curvearrowright$	Left turn of vector, right turn of vector

1

Locating ROI in Iris Using Randomized Hough Transform

Paras Gulati[1] and Teekam Singh[2]

[1]Department of Computer Science & Engineering, Graphic Era Hill University, Dehradun-248002, India
[2]Department of Mathematics, Graphic Era Hill University, Dehradun-248002, India,
E-mail: parasgulati150@gmail.com; tsingh@gehu.ac.in; tsingh@ma.iitr.ac.in

Abstract

Iris segmentation is applied to the human eye to extracting the region of interest (ROI) and is used to recognize the human uniquely and efficiently. In this, we have first detected and removed the reflection mask using adaptive thresholding; then to detect edges, the Gaussian filter of Canny edge detection is used, and then the resultant reflection and boundary mask free image is used to locate the approximate iris center using randomized Hough transform and then the polar transform is used to find ROI.

Keywords: Image segmentation, iris, iris recognition, Hough transforms.

1.1 Introduction

Iris-based identification of humans has gained significant attention due to its authenticity [1]. In this identification, the iris is uniquely identified by their patterns which are unique for everyone. These patterns are stable over the decades. But recognition of region of interest (ROI) in iris can be affected due to the poor quality of image and noise present in the image. So, to overcome these problems, a systematic and vigorous segmentation of these images is needed. Andreas Uhl and Peter Wild [2] proposed an iris segmentation algorithm that consists of two stages. They used an adaptive Hough transform (AHT) for detecting center imprecisely and they used the transformation of polar coordinates for detecting boundary which produces better results over the agent-based center localization method. We used the randomized Hough transform which makes use of geometric properties of analytical curves rather than using normal Hough transform which works upon voting mechanism for every nonzero pixel in the image. It performs efficiently in terms of time and space of the process.

The biometric feature such as iris retina and palm lines serve as a kind of living passport that everyone carries with them everywhere they go [3]. Iris-based identification was evolved in 1992, and, since then, its usage is increasing heavily. When we use Hough transformation to detect edges and closed iris areas, the result achieved is robust to noise and interval of curves, but the major disadvantage in Hough transformation is that calculating quantity is very large; so with the increase in image size, the processing time also increases [4]. So to alleviate this problem, randomized Hough transformation is used. Randomized Hough transformation efficiently detects global maximum pixel values in the accumulator space [5]. The global maximum point (a, b) which describes the detected curve can be removed from the image for applying algorithm again for the remaining pixels.

1.2 Dataset Used

CASIA-Iris Syn subset of CASIAV4 iris data of Centre for Biometrics and Security Research is used as the dataset. The dataset contains 10,000 iris images from 1000 classes. Each image is 640 × 480 pixels. It contains intraclass variations due to eyeglasses and specular reflection. We have used and tested our algorithm on eyeglasses and eyebrow classes of the dataset. It contains 100 images each for eyeglasses and eyebrow classes.

CASIA-Iris V4 dataset contains 54,607 images of iris in total. Each of those images is a gray-level lk JPEG images. In each of this image, a different level of characteristic of iris is captured, which provides a robust field for research in segmentation of Iris images. Each of these images is captured under infrared illumination.

CASIA-Iris V3 dataset has three subsets: CASIA-Iris-Interval, CASIA-Iris-Lamp, and CASIA-Iris-Twins. CASIA-Iris has V3 and V4 datasets out of which we have used CASIA-Iris V4 for applying our segmentation algorithm.

1.3 Segmentation Procedure

The technique of extracting the target region out of an image by maintaining the quality of an image is called segmentation [6]. The most crucial stage in this whole process of iris recognition is iris segmentation. In this paper, an algorithm of iris segmentation is presented which is designed based on the natural properties of the iris under general conditions.

The first and the most important step in this algorithm is iris segmentation. The main objective of iris segmentation is to draw out the major distinguishing patterns in iris, for which we need to separate out noise and other minor patterns in iris. If we are able to find these major distinguishing patterns accurately, then the accuracy of uniquely recognizing using iris also increases exponentially [7].

1.3.1 Reflection Mask Detection and Removal

Reflection is enough as noise to deviate the machine to end up making wrong decisions; any reflection in pupil can cause a problem in the proper location of the ROI. So, for the detection of any reflection, adaptive thresholding is used [8].

1.3.2 Thresholding

Thresholding is a technique which is used to segmentation image into contours by separating different pixel intensity values with the help of a threshold [9], say

$$pixel\_value\_1 = \left\{ \begin{array}{l} x\ if(pixel\_value\_0 < threshold) \\ \\ y\ if(pixel\_value\_0 > threshold) \end{array} \right\}$$

Thresholding is of the following types: point thresholding, region thresholding, and global thresholding.

The thresholding that we have used, i.e., adaptive thresholding, is region thresholding.

1.3.3 Adaptive Thresholding

Image thresholding is a technique that is being used for decreasing the unique values of pixels in an image. When we impractically separate the pixels based on a particular value, then it might create problems in certain cases, and for overcoming those problems, adaptive thresholding is discovered. In adaptive thresholding, the value that we used in normal thresholding is here selected at run time, by the algorithm according to its spatial distribution of pixels.

Adaptive thresholding segments the image into contours. A region around a pixel is selected *M (b X b)*. It selects all pixels with intensity (*H*) more than the local mean of that region [10].

$$H_{i,j} = \frac{1}{2\pi\sigma^2} \exp\left(-\frac{(i-(k+1))^2 + (j-(k+1))^2}{2\sigma^2}\right)$$

$$1 \leq i, j \leq (2k+1)$$

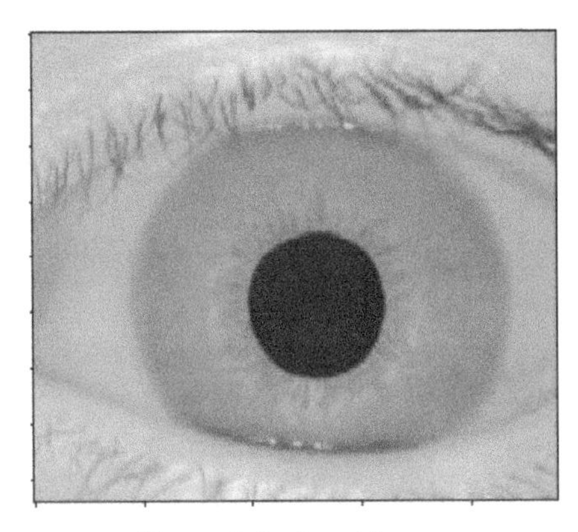

Figure 1.1 Input image.

Figure 1.2 Mask detected.

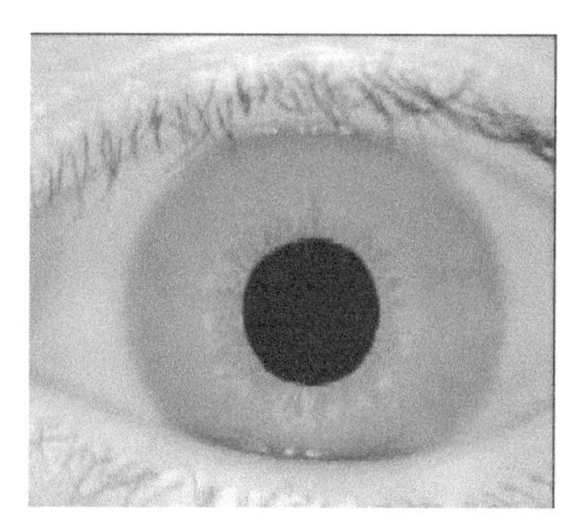

Figure 1.3 Mask removed.

1.3.4 Edge and Border Mask Detection

Edge is the most critical part of image segmentation; a minute change in intensity value at edges can make a significant effect on the unique identification of iris [11].

When there is a change of intensity in a digital image, then it is known as edge. It is a sudden change of discontinuities, noticed in an image. The three types of edges are horizontal, vertical, and diagonal edges.

To segment the non-trivial images that seem to be sometimes complicated, some factors that need to be taken into consideration are edge-like features, noise, image brightness, and corners. Image segmentation is based on two major characteristic features of intensity values which are (i) discontinuity, which involves the partitioning of an image based on sudden changes in intensity, and (ii) similarity, which involves the partitioning of an image area that look alike or similar based on some predefined criteria [12].

1.3.5 Edge Detector of Canny

For edge detection, the major techniques used are Sobel and Canny detectors. In Sobel detector, it uses a filter of size 3×3 which finds edge features and recognizes it. But the Canny detector is little advanced compared to Sobel detector; it removes the noise before recognizing the edges.

For detecting a range of edges in an image, the Canny detection algorithm has few intermediate stages. Fundamental concepts of Canny edge detector are [13, 14].

Canny detector works according to five steps: smoothing, finding, gradients, non-maximum suppression, double thresholding, edge tracking by hysteresis.

1) Every edge in the image is very crucial; so it should not be missed, and no extra edge (edge not in image) should be detected by the detector.
2) Edge points are well localized; so the detected edge should be almost near about the actual edge.
3) A single edge should be detected only once.

1.3.6 Working of Canny Edge Detector

The Canny edge detection algorithm was proposed by John F. Canny. It is the enhanced version of Sobel edge detection algorithm. It considers three major criteria. The most crucial is to detect all the important edges in the image. The second is the concept of localization; in this, the detected edge point should be as close as possible to the true edge. The third is, for each edge, only one response should be entertained [15].

The smoothening of image performed by the detector eliminated noise. Then image gradients are used to highlight the regions with greater spatial derivative. Then it performs non-maximum suppression along the found regions, and to track the non-suppressed pixels, hysteresis is used.

1.3.7 Gaussian Mask (Filter)

It smoothed the image by removing noise. The Gaussian mask is convolved with the image [16].

$$H_{i,j} = \frac{1}{2\pi\sigma^2} \exp\left(-\frac{(i-(k+1))^2 + (j-(k+1))^2}{2\sigma^2}\right)$$

$$1 \leq i,j \leq (2k+1),$$

where $H_{i,j}$ is Gaussian mask for kernel $= (2k+1) \times (2k+1)$.

To decrease the sensitivity towards the noise, the width of the mask is increased. But with the increase of mask width, localization error is also increased.

Gaussian kernel ($H_{i,j}$) size affects the performance of Canny detector to a greater extend. Experiments show that greater the size of kernel, the lesser its sensitivity to noise.

The size of Gaussian kernel highly affects the performance of the detector. The localization error also increases with the increase of kernel size. Experimentally, it has been shown that kernel size of 5×5 is well suited for the iris segmentation.

1.3.8 Intensity Gradient

Sobel Feldman operator is used to perform measurement of 2D spatial gradient on image [17], after which the approximate strength of the edge is found. The Sobel Feldman operator performs measurement of a 2D spatial gradient on the image [18], and then for each pixel point in the image, the approximate edge strength is found. The Feldman operator uses two convolution masks of dimensions 3×3, out of which the first calculates the gradients across x-direction and the second calculates the gradient across y-direction.

$$G_x = \begin{pmatrix} -1 & 0 & 1 \\ -2 & 0 & 2 \\ -1 & 0 & 1 \end{pmatrix}, \quad G_y = \begin{pmatrix} 1 & 2 & 1 \\ 0 & 0 & 0 \\ -1 & -2 & -1 \end{pmatrix}$$

$$|G| = |G_x| + |G_y|,$$

$$\theta(x,y) = \tan^{-1}\left(\frac{G_y}{G_x}\right),$$

where G is the magnitude and θ is the slope. Once the edge directions are calculated, the edge directions are mapped to the direction that can be traced in the image.

1.3.9 Non-Maximum Suppression

Non-maximum suppression is applied along the edges; in the direction of edges [19], this restrains pixel value that is not inspected in the edge and has non-maximum intensity pixels.

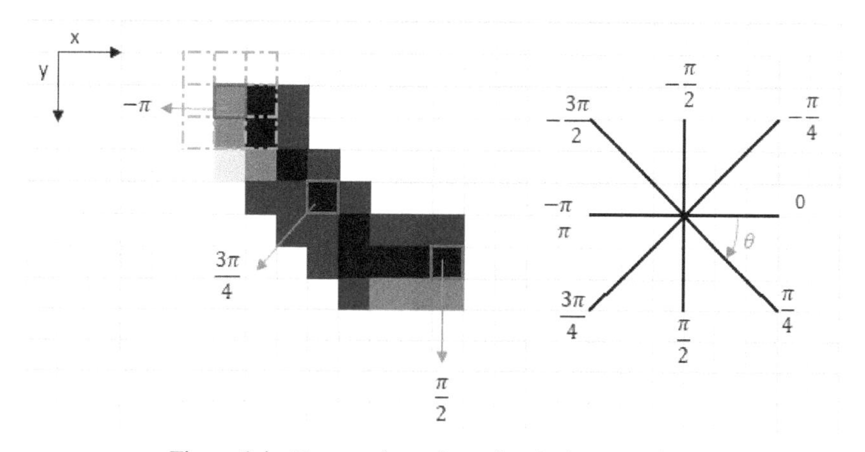

Figure 1.4 Non-maximum intensity pixel suppression.

It finds the pixel which has maximum value at an edge. It uses comparison between different pixel values; for each pixel value, it is compared with the two neighboring pixels in the gradient direction, and if the corresponding pixel value is greater than its neighboring pixel values, then its value is retained, else, it is set to zero.

1.3.10 Edge Tracking by Hysterics

Hysterics is used for excluding streaks [20]. Streaking is the method in which we break edge contours that are caused by the above fluctuating operator. To avoid this, hysterics uses two threshold values: max (high) and min (low)

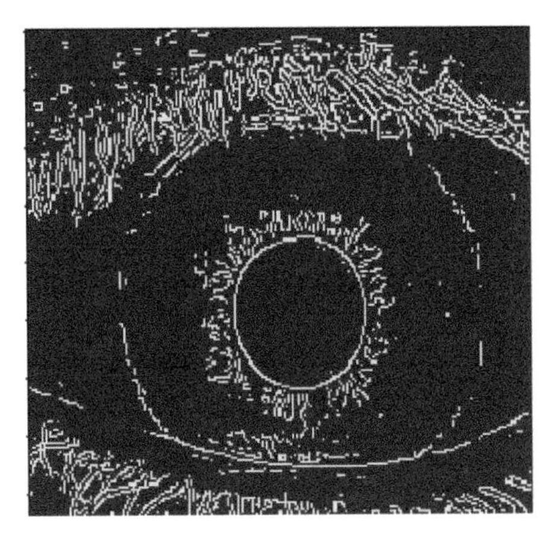

Figure 1.5 Edge and boundary mask detection.

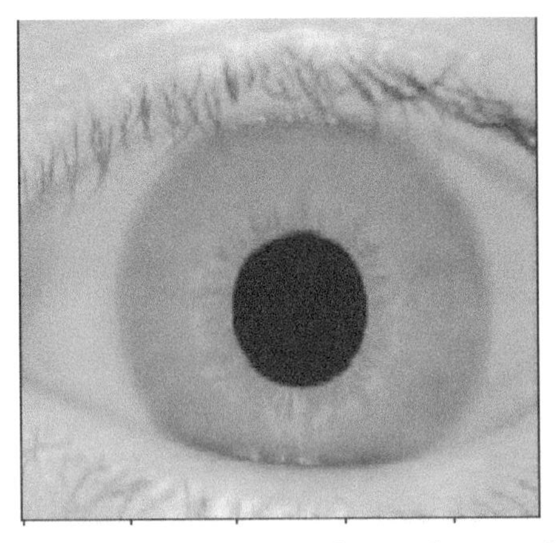

Figure 1.6 Edge and boundary mask removed.

(T1 and T2). Every pixel in the image is checked, and, if found greater than T1, then it is assumed to be an edge pixel and any pixel connected to the founded edge pixel and having a value greater than T2 is also assumed to be an edge pixel.

To get an accurate result, regions with the weak edges are removed. Due to this noise, response edges are unconnected and only true edges be it caused by weak edge pixel or strong edge pixel will remain connected.

1.4 Classical Hough Transform

Features are the unique representation of traits in the iris. And one of the techniques used for feature extraction is Hough transformation. It polls all the objects falling in various classes, and this helps it find the imperfect instances of such objects [21].

The classical Hough transformation was designed for the unique identification of lines in the image, but with the advancements in the digital image processing, it was extended for identification of various shapes like circles, lines, etc.

1.4.1 Transformation for Circle Detection

To detect a circle or circular arc using fast randomized Hough transform, one point is picked at random as a seed point. Then we check according to rules whether the picked point lies inside or outside of the circle.

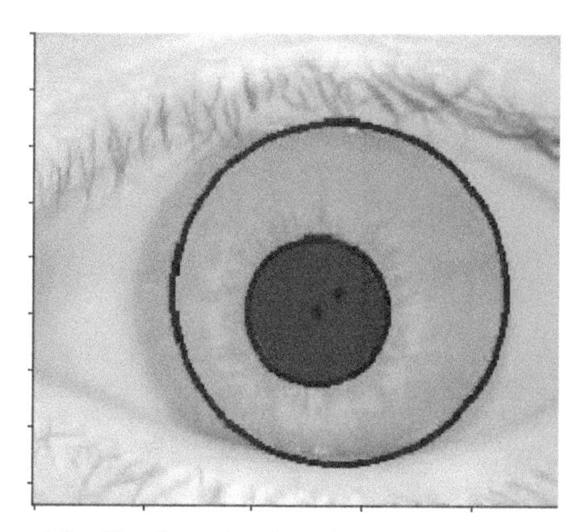

Figure 1.7 Circular region detection using Hough transform.

For applying a Hough transform on circular objects, a certain parametric equation is used; a circle can be represented as

$$(x - ax)^2 + (y - ay)^2 = R^2,$$

where (ax, ay) denotes center of the circle and *R* denotes radius. The process of circle detection is done in two steps. 2 This process can be divided into two steps.

1. In this step, the optimal center of circle is found by fixing the radius in 2D parameter space.
2. Find the optimal radius in 1D parameter space.

1.4.2 Polar Transform

For locating any point on the 2D plane, we have Cartesian system, in which we specify *x* and *y* coordinates from origin. But there is one more way of locating any point on 2D plane, and that is polar coordinates. In this, any point $(x, y$ is represented as radius (*r*), distance from origin, and the angle it subtends from *r* and the positive *x*-axis.

For any given point represented as (r, θ) can be represented uniquely by adding 360 to θ any number of times.

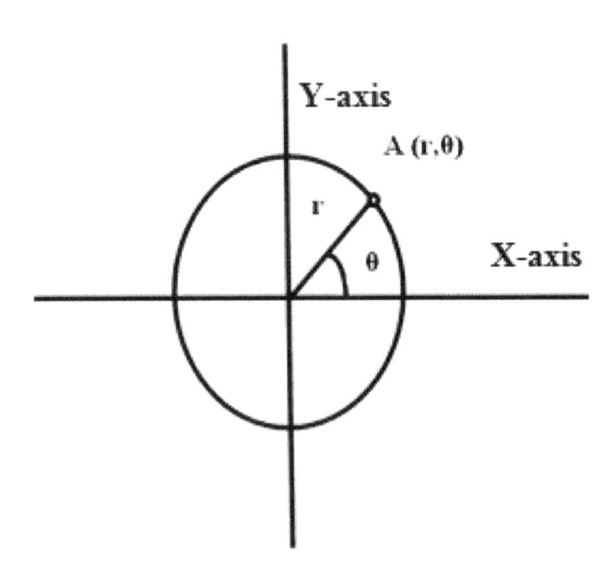

Figure 1.8 Polar coordinates.

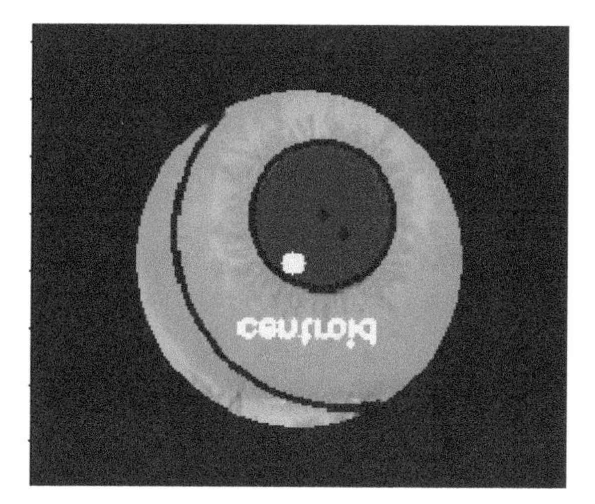

Figure 1.9 Iris extraction using polar transform.

And in polar coordinate system, to reverse the direction of point, we add 180 to θ.

Polar transformation is changing orientation of polar coordinates. In polar representation, a same point represented as r and θ can be represented by infinite number of ways.

Polar System: it is used to represent a point in the system with the help of its distance from center r and its angle θ from the x-axis.

The conversion of a system from the Cartesian coordinate system into a polar system is called a polar transform [22]

$$\begin{aligned} x &= r.\cos(\theta) \\ y &= r.\sin(\theta) \\ r^2 &= x^2 + y^2 \\ \theta &= \tan^{-1}\left\{\frac{y}{x}\right\}, \end{aligned}$$

where r is the radius of the circle and θ is the angle between the line joining the point to the pole and the x-axis.

Log-polar transform (LPT): It is a tool which is used to rotate and scale the images. But while using it with altered images, it gives disastrous results [22].

1.5 Conclusion

In general, iris segmentation is built according to a particular database. But this leads to the lack of scalability. In this paper, we have proposed a two-stage iris segmentation framework. The segmentation algorithms are measured or compared based on the three main factors also known as quality factors, accuracy, usability, and speed. The proposed two-stage algorithm requires less processing time. It provides a robust method for the iris-based image segmentation. Upon experimenting, we found that the proposed method is scalable, and it does not depend on the iris database used. Hence, the proposed method of randomized Hough transform increases efficiency both in terms of time and space as compared to existing database-dependent iris segmentation procedures.

References

[1] M. Nabti, A. Bouridane, 'An improved iris recognition system using feature extraction based on wavelet maxima moment invariants', Proc. of International Conference on Biometric, pp. 988–996, Springer-Berlin, Heidelberg, 2007.

[2] U. Andreas, W. Peter, 'Weighted adaptive Hough and ellipsopolar transform for real-time iris segmentation', Proc. of the Fifth International Conference on Biometric, pp. 283–290, IEEE, 2012.

[3] P. P. Polash, M. M. Monwar, 'Human iris recognition for biometric identification', 10th Proc. of International Conference on Computer and Information Technology, Dhaka, pp. 01–05, 2007.

[4] J. Illingworth, J. Kittler, 'A survey of the Hough transform', Computer Vision, Graphics, and Image Processing, vol. 44, no. 1, pp. 87–116, 1988.

[5] P. Kultanen, L. Xu, E. Oja, 'A new curve detection method: randomized Hough transform (RHT)', Patterns Recognition Letters, vol. 11, no. 5, pp. 331–338, 1990.

[6] X. Yuan, S. Pengfei, 'An iris segmentation procedure for iris recognition', Proc. of the Chinese Conference on Biometric Recognition. Springer, Berlin, Heidelberg, pp. 546–553, 2004.

[7] A. Bendale, A. Nigam, S. Prakash, P. Gupta, 'Iris segmentation using improved hough transform', International Conference on Intelligent Computing, Springer, Berlin, Heidelberg, pp. 408–415, 2012.

[8] F. Fuentes-Hurtado, V. Naranjo, J. A. Diego-Mas, M. A. Alcañiz, 'A hybrid method for accurate iris segmentation on at-a-distance visible-wavelength images', EURASIP Journal on Image and Video Processing, vol. 2019, no. 1, pp. 75, 2019.
[9] R. López-Leyva et al., 'Comparing threshold-selection methods for image segmentation: application to defect detection in automated visual inspection systems', Mexican Conference on Pattern Recognition, pp. 33–43, Springer, Cham, 2016.
[10] P. Roy et al., 'Adaptive thresholding: a comparative study', International Conference on Control, Instrumentation, Communication and Computational Technologies (ICCICCT), pp. 1182–1186, IEEE, 2014.
[11] S. Wang, G. Feng, L. Tiecheng, 'Evaluating edge detection through boundary detection', EURASIP Journal on Advances in Signal Processing, vol. 2006, no. 1, pp. 076278–076292, 2006.
[12] J. S. Owotogbe, T. S. Ibiyemi, B. A. Adu, 'Edge Detection Techniques on Digital Images-A Review', International Journal of Innovative Science and Research Technology, vol. 4, no. 1, pp. 329–332, 2019.
[13] J. Li, D. Sheng, 'A research on improved canny edge detection algorithm', International Conference on Applied Informatics and Communication, Springer, Berlin, Heidelberg, pp. 102–108,2011.
[14] C. Gentsos et al., 'Real-time canny edge detection parallel implementation for FPGAs', 17th IEEE International Conference on Electronics, Circuits and Systems, IEEE, pp. 499–502, 2010.
[15] R. Muthukrishnan, R. Miyilsamy, 'Edge detection techniques for image segmentation', International Journal of Computer Science & Information Technology, vol. 3, no. 6, pp. 259–271, 2011.
[16] E. S. Gedraite, M. Hadad, 'Investigation on the effect of a Gaussian Blur in image filtering and segmentation', Proc. ELMAR-2011, IEEE, pp. 393–396, 2011.
[17] J. A. Saif, M. H. Hammad, I. A. Alqubati, 'Gradient based image edge detection', International Journal of Engineering and Technology, vol. 8, no. 3, pp. 153–156, 2016.
[18] A. Eshaghzadeh, N. Salehyan, 'Canny edge detection algorithm application for analysis of the potential field map', Geophysics, vol. 62, no. 2016, pp. 807–813, 2016.
[19] R. Medina-Carnicer et al., 'On candidates' selection for hysteresis thresholds in edge detection', Pattern Recognition, vol. 42, no. 7, pp. 1284–1296, 2009.

[20] M. Rizon et al., 'Object detection using circular Hough transform', American Journal of Applied Sciences, vol. 2, no. 1, pp. 1606–1609, 2005.

[21] A. S. Hassanein, M. Sherien, M. Sameer, M. E. Ragab, 'A survey on Hough transform, theory, techniques and applications', arXiv preprint arXiv: 1502.02160, 2015.

[22] R. Matungka, F. Z. Yuan, L. E. Robert, 'Image registration using adaptive polar transform', IEEE Transactions on Image Processing, vol. 18, no. 10, pp. 2340–2354, 2009.

2

Cross-Diffusion, Nonlinear Radiation, and Newtonian Heating Effects of Third-Grade Fluid Flow Over a Riga Plate With Entropy Generation Minimization

K. Loganathan[1], M. Mohanraj[2] and K. Tamilvanan[3]

[1]Research and Development Wing, Live4Research, Tiruppur, Tamil Nadu 638106, India
[2]Department of Mechanical Engineering, Hindusthan College of Engineering and Technology, Coimbatore, Tamil Nadu 641032, India
[3]Department of Mathematics, Government Arts College for Men, Krishnagiri, Tamil Nadu 635 001, India
E-mail: loganathankaruppusamy304@gmail.com

Abstract

In this chapter, we investigate the entropy generation of third-grade liquid flow induced by a Riga plate. Cross-diffusion, nonlinear radiation, and Newtonian heating effects are incorporated into the physical model. Additionally, the entropy generation of third-grade liquid is calculated. The governing system is simplified into the nonlinear ordinary system from partial differential equations. The resulting nonlinear ordinary differential equations are computed through the homotopy analysis method (HAM). The effects of all evolving parameters are examined with the aid of graphs and tables. We noticed that there is an increase in velocity but a diminishing phenomenon in concentration, temperature, and entropy profiles with the swelling in the range of modified Hartmann number.

Keywords: Third-grade fluid, Riga plate, nonlinear radiation, Soret and Dufour effects, HAM technique.

2.1 Introduction

The second law of thermodynamics and its application of the thermal engineering system are examined by several researchers in recent years. Rashidi *et al.* [1] explored the entropy effects of electrically conducting fluid toward a rotating disk with a uniform magnetic field. Governing equations are solved by a homotopy scheme and then utilized artificial neural network and particle swarm optimization procedure to minimize the entropy generation. Malvandi *et al.* [2] presented the fluid flow toward a flat sheet with heat transfer effects through Runge–Kutta–Fehlberg scheme. Entropy generation minimization is applied to attain a thermodynamic optimization. Hedayati *et al.* [3] discussed the entropy optimization of fluid flow with heat transfer caused by a wedge. Rashidi *et al.* [4] inspected the electrically conducting nanofluid flowing toward a rotating porous disk with the analysis of the second law of thermodynamics. The latest developments in entropy minimization were examined in previous research works [5–8].

The mathematical modeling which represents physical systems does not have the exact solutions in the form of strongly nonlinear equations. The analytical or numerical techniques can be utilized to resolve these nonlinear equations, for instance, Adomian decomposition method (ADM) [9], differential transform method (DTM) [10], Runge–Kutta method [11], and the present employed analytical method, HAM [12–15].

The present investigation is mainly inspired by the need to know the entropy minimization analysis for third-grade liquid flow caused by a Riga plate with Newtonian surface boundaries and Christov–Cattaneo heat flux conditions. HAM scheme was used for resolving the governing ODE and correspondingly studying the impacts of velocity, nanoparticle volume fraction, temperature, entropy generation, and Bejan number. The engineering instrument performances reduce in the attendance of the irreversibility and subsequently the entropy generation is a way of calculating the process irreversibility. Meanwhile, the non-Newtonian fluid flow and heat transmission signify several imperative usages, for instance, artificial fibers and plastic films.

2.2 Problem Development

We are attracted to examine the entropy analysis of third-grade fluid flow over a Riga plate with cross-diffusion, nonlinear radiation, and Newtonian heating effects. A physical diagram representing the coordinate system along with the formation of the model is shown in Figure 2.1. The governing physical systems are stated as

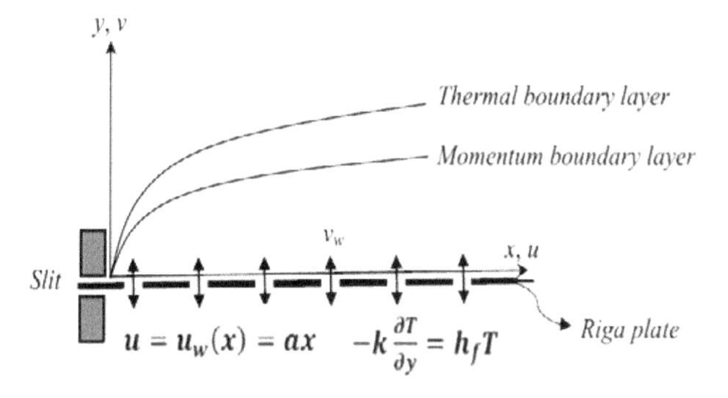

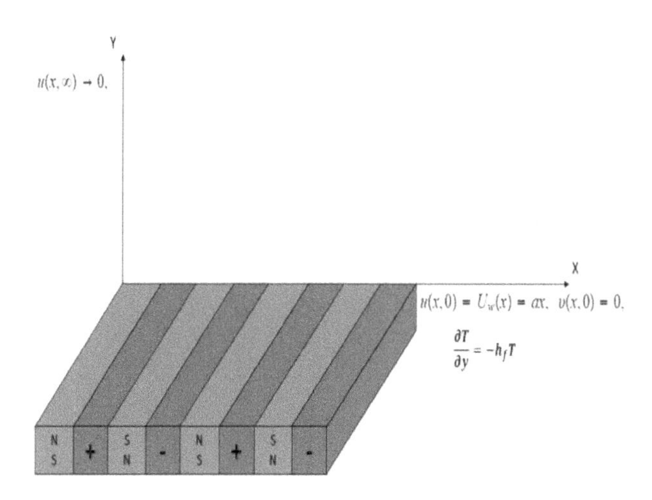

Figure 2.1 Schematic diagram.

$$\frac{\partial u}{\partial x} + \frac{\partial v}{\partial y} = 0 \tag{2.1}$$

$$u\frac{\partial u}{\partial x} + v\frac{\partial u}{\partial y} = \nu\frac{\partial^2 u}{\partial y^2} + \frac{A_1^*}{\rho}\left(u\frac{\partial^3 u}{\partial y^2 \partial x} + v\frac{\partial^3 u}{\partial y^3} + \frac{\partial u}{\partial x}\frac{\partial^2 u}{\partial y^2} + 3\frac{\partial u}{\partial y}\frac{\partial^2 u}{\partial x \partial y}\right) +$$
$$2\frac{A_2^*}{\rho}\frac{\partial u}{\partial y}\frac{\partial^2 u}{\partial x \partial y} + 6\frac{\beta_1^*}{\rho}\left(\frac{\partial u}{\partial y}\right)^2\frac{\partial^2 u}{\partial y^2} + \frac{\pi j_0 M_0 exp\left(-\frac{\pi}{b}y\right)}{8\rho} \tag{2.2}$$

$$u\frac{\partial T}{\partial x}+v\frac{\partial T}{\partial y}+\lambda_T\left(u^2\frac{\partial^2 T}{\partial x^2}+v^2\frac{\partial^2 T}{\partial y^2}+\left(u\frac{\partial u}{\partial x}\frac{\partial T}{\partial x}+v\frac{\partial u}{\partial y}\frac{\partial T}{\partial x}\right)\right.$$
$$\left.+2uv\frac{\partial T^2}{\partial x\partial y}\right)+\left(u\frac{\partial v}{\partial x}\frac{\partial T}{\partial y}+v\frac{\partial v}{\partial y}\frac{\partial T}{\partial y}\right)=\frac{k}{\rho c_p}\frac{\partial^2 T}{\partial y^2}+\frac{1}{\rho c_p}\frac{16\sigma^*}{3k^*}$$
$$\left[3T^2\left(\frac{\partial T}{\partial y}\right)^2+T^3\frac{\partial^2 T}{\partial y^2}\right]+\frac{Q_0}{\rho c_p}\left(T-T_\infty\right)+\frac{D_m k_T}{C_s C_p}\frac{\partial^2 C}{\partial y^2} \tag{2.3}$$

$$u\frac{\partial C}{\partial x}+v\frac{\partial C}{\partial y}=D_m\frac{\partial^2 C}{\partial y^2}-k_m\left(C-C_\infty\right)+\frac{D_m k_T}{T_m}\frac{\partial^2 T}{\partial y^2}. \tag{2.4}$$

Boundary points are

$$u=u_w\left(x\right)=ax,\; v=0,\; -k\frac{\partial T}{\partial y}=h_f T\; c=c_w\; at\; y=0$$
$$u\to 0,\; v\to 0,\; T\to T_\infty,\; c\to c_\infty\;\; as\; y\to\infty. \tag{2.5}$$

Consider the similarity transformation given below

$$\psi=\;\sqrt{a\nu}xF_1\left(\zeta\right),\; u=\frac{\partial\psi}{\partial y},\; v=\;-\frac{\partial\psi}{\partial x},\;\; \zeta=\sqrt{\frac{a}{\nu}}y,\; u=axF_1^{'}\left(\zeta\right),$$
$$v=-\sqrt{a\nu}F_1\left(\zeta\right),\theta_1\left(\;\zeta\right)=\frac{T-T_\infty}{T_\infty},\; \phi_1\left(\zeta\right)=\frac{C-C_\infty}{C_w-C_\infty}. \tag{2.6}$$

The governing equations are given by the following:

$$F_1^{'''}-F_1^{'2}+F_1F_1^{''}+\alpha_1\left(2F_1^{'}F_1^{'''}-F_1F_1^{iv}\right)+\left(3\alpha_1+2\alpha_2\right)F_1^{''2}+$$
$$6\beta_1 ReF_1^{'''}F_1^{''2}+Qe^{-d_1\zeta}=0 \tag{2.7}$$

$$\frac{1}{Pr}\theta_1^{''}+\frac{Rd}{Pr}((\theta_1\left(\theta_w-1\right)+1))^2\left(3\theta_1^{'2}\left(\theta_w-1\right)+\left(\theta_1\left(\theta_w-1\right)+1\right)\theta_1^{''}\right)+$$
$$F_1\theta_1^{'}+H_g\theta_1-\gamma\left(F_1^2\theta_1^{\phi}+F_1F_1^{'}\theta_1^{'}\right)+D_F\phi_1^{''}=0 \tag{2.8}$$

$$\frac{1}{S_C}\phi_1^{''}+F_1\phi_1^{'}-C_r\phi_1+S_r\theta_1^{''}=0. \tag{2.9}$$

Boundary points are

$$F_1\left(0\right)=0,\; F_1^{'}\left(\infty\right)\to\; 0,\; F_1^{'}\left(0\right)=1,\theta_1^{'}\left(0\right)=-Nw\left(1+\theta_1\left(0\right)\right),$$
$$\theta_1\left(\infty\right)\to\; 0,\; \phi_1\left(0\right)=1,\; \phi_1\left(\infty\right)\;\to\; 0. \tag{2.10}$$

The dimensionless form becomes

$$\alpha_1 = \frac{aA_1^*}{\nu}, \alpha_2 = \frac{aA_2^*}{\nu}, \beta_1 = \frac{a\beta_1^*}{\nu}, Re = \frac{u_w x}{\nu}, Q = \frac{\pi j_0 M_0}{8a^3 x\rho},$$
$$Rd = \frac{\left(16\sigma^* T_\infty^3\right)}{3kk^*}, \theta_w = \frac{T_w}{T_\infty}, Pr = \frac{\rho C_p}{k}, H_g = \frac{Q_0}{\rho c_p}, \gamma = \lambda_T a,$$
$$D_F = \frac{D_m k_T}{\nu C_s C_p}\frac{C_w - C_\infty}{T_\infty}, S_r = \frac{D_m k_T}{\nu T_m}\frac{T_\infty}{C_w - C_\infty}, S_C = \frac{\nu}{D_m}, C_r = \frac{k_m}{a}.$$

The physical entitles are defined as follows:

$$Re^{\frac{1}{2}} C_{F_{1x}} = F_1^{''}(0) + \alpha_1 F_1^{'}(0) F_1^{'''}(0) + \beta_1 Re[F_1^{''}(0)]^3$$
$$Re^{-\frac{1}{2}} Nu_x = -(1 + \frac{3}{4} Rd \left((\theta_w - 1)(\theta_1(0))^3 \right))\theta_1^{'}(0)$$
$$Re^{-\frac{1}{2}} Sh_x = -\phi_1^{'}(0).$$

2.3 Entropy Optimization

The entropy rate per unit volume for the third-grade liquid is as follows:

$$S_{gen}^{'''} = \frac{K_1}{T_\infty^2}\left[\left(\frac{\partial T}{\partial x}\right)^2 + \left(\frac{\partial T}{\partial y}\right)^2 + \frac{16\sigma^* T_\infty^3}{3kk^*}\left(\frac{\partial T}{\partial y}\right)^2\right] +$$
$$\frac{\mu}{T_\infty}\left[2\left(\frac{\partial u}{\partial x}\right)^2 + \left(\frac{\partial v}{\partial y}\right)^2\right] + \left[\frac{\partial u}{\partial y} + \frac{\partial v}{\partial x}\right]^2 + \frac{RD}{C_\infty}\left[\left(\frac{\partial C}{\partial x}\right)^2 + \left(\frac{\partial C}{\partial y}\right)^2\right] +$$
$$\frac{RD}{T_\infty}\left[\left(\frac{\partial T}{\partial x}\right)\left(\frac{\partial C}{\partial x}\right) + \left(\frac{\partial T}{\partial y}\right)\left(\frac{\partial C}{\partial y}\right)\right]. \tag{2.11}$$

By applying the boundary-layer approximation, Equation (2.11) is modified as

$$S_{gen}^{'''} = \frac{K_1}{T_\infty^2}\left[\left(\frac{\partial T}{\partial y}\right)^2 + \frac{16\sigma^* T_\infty^3}{3kk^*}\left(\frac{\partial T}{\partial y}\right)^2\right] + \frac{\mu}{T_\infty}\left(\frac{\partial u}{\partial y}\right)^2 +$$
$$\frac{RD}{C_\infty}\left(\frac{\partial C}{\partial y}\right)^2 + \frac{RD}{T_\infty}\left(\frac{\partial T}{\partial y}\right)\left(\frac{\partial C}{\partial y}\right). \tag{2.12}$$

Characteristic entropy generation rate declared as $S_0^{'''}$

$$S_0^{'''} = \frac{K_1}{T_\infty^2}\frac{(\triangle T)^2}{l^2}. \tag{2.13}$$

The dimensionless structure of entropy generation is calculated through

$$E_G = \frac{S'''_{gen}}{S'''_0}. \tag{2.14}$$

The dimensionless form of entropy generation number stated as follows:

$$E_G = Re\left(1 + Rd\left(1 + (\theta_1(\theta_w - 1) + 1)^2\right)\right)\theta_1'^2 + Re\frac{Br}{\Omega}F_1''^2 + Re\left(\frac{\varepsilon}{\Omega}\right)^2\lambda\phi_1'^2 + Re\frac{\varepsilon}{\Omega}\lambda\phi_1'\theta_1'. \tag{2.15}$$

Bejan number mathematically exposed as ratio between heat and mass transfer and total entropy rate

$$\mathrm{Be} = \frac{Re\left(1 + Rd\left(1 + (\theta_1(\theta_w - 1) + 1)^2\right)\right)\theta_1'^2 + Re\left(\frac{\zeta}{\Omega}\right)^2\lambda\phi_1'^2 + Re\frac{\zeta}{\Omega}\lambda\phi_1'\theta_1'}{\left[Re\left(1 + Rd\left(1 + (\theta_1(\theta_w - 1) + 1)^2\right)\right)\theta_1'^2 + Re\frac{Br}{\Omega}F_1''^2 + Re\left(\frac{\zeta}{\Omega}\right)^2\lambda\phi_1'^2 + Re\frac{\zeta}{\Omega}\lambda\phi_1'\theta_1'\right]}. \tag{2.16}$$

2.4 Solution Technique (HAM)

The primary assumptions of homotopy technique is

$$F_{10}(\zeta) = 1 - Exp(-\zeta), \theta_{10}(\zeta) = \frac{Nw * Exp(-\zeta)}{1 - Nw}, \phi_{10} = Exp(-\zeta).$$

The auxiliary linear operators L_{F_1}, L_{θ_1}, and L_{ϕ_1} are derived as

$$L_{F_1} = F_1^{\phi}(\zeta) - F_1'(\zeta)$$

$$L_{\theta_1} = \theta_1^{\phi}(\zeta) - \theta_1(\zeta)$$

$$L_{\phi_1} = \phi_1^{\phi}(\zeta) - \phi_1(\zeta)$$

$$L_{F_1}\left[E_1 + E_2 e^{\zeta} + E_3 e^{-\zeta}\right] = 0$$

$$L_{\theta_1}\left[E_4 e^{\zeta} + E_5 e^{-\zeta}\right] = 0$$

$$L_{\phi_1}\left[E_6 e^{\zeta} + E_7 e^{-\zeta}\right].$$

The special solutions $[F_{1m}^{*}, \theta_{1m}^{*}, \phi_{1m}^{*}]$ are

$$F_{1m}(\zeta) = F_{1m}^{*}(\zeta) + E_1 + E_2 e^{\zeta} + E_3 e^{-\zeta},$$
$$\theta_{1m}(\zeta) = \theta_{1m}^{*}(\zeta) + E_4 e^{\zeta} + E_5 e^{-\zeta},$$
$$\phi_{1m}(\zeta) = \phi_{1m}^{*}(\zeta) + E_6 e^{\zeta} + E_7 e^{-\zeta}$$

where $E_j\ (j = 1 - 7)$ denotes the arbitrary conditions.

The non-dimensional constraints are fixed at $\alpha_1 = \alpha_2 = \beta_1 = Re = C_r = Nw = d_1 = 0.1$, $Pr = 0.9$, $S_C = 0.9$, $Rd = 0.3$, $\theta_w = 0.5$, $H_g = -0.3$, $D_F = 0.5$, and $S_r = 0.3$ unless we state that. Figure 2.2 represents the convergence range of parameters h_{F_1}, h_{θ_1}, and h_{ϕ_1}. From this figure, we obtain a convergence control range of total domain is $h_{F_1} = h_{\theta_1} = h_{\phi_1} = -0.7$. Table 2.1 depicts the order of approximations of HAM up to five decimal places.

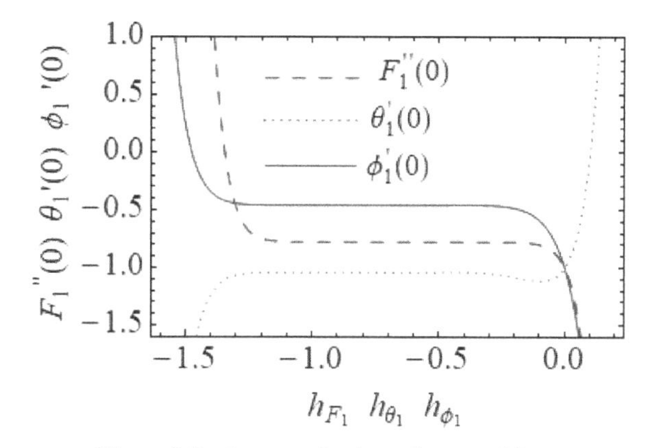

Figure 2.2 h-curves for h_{F_1}, h_{θ_1}, and h_{ϕ_1}.

Table 2.1 Order of approximations of HAM.

Order of Approximations	$-F_1''(0)$	$-\theta_1'(0)$	$-\phi_1'(0)$
1	0.77950	1.15750	0.65777
5	0.77574	1.05662	0.45597
10	0.77594	1.03490	0.45666
15	0.77593	1.04097	0.45780
20	0.77593	1.03986	0.45730
25	0.77593	1.03986	0.45741
30	0.77593	1.03995	0.45741
35	0.77593	1.03995	0.45741
40	0.77593	1.03995	0.45741

2.5 Computational Results and Discussion

In this section, we express the results of third-grade fluid flow with nonlinear radiation, cross-diffusion, and Newtonian heating effects. The ordinary governing systems were solved with the assistance of HAM. All the emerging parameters are discussed and reported in detail.

The effects of modified Hartmann number (Q) for various profiles are noticed in Figure 2.3(a)–(e). From Figure 2.3(a), we noted that the velocity profile $\left(F_1^{'}\right)$ enhances for ascending values of Q; this factor is due to Lorentz force. The opposite trends are noted in temperature profile (θ_1), concentration profile (ϕ_1), and entropy profile (E_G) for larger values of modified Hartmann number (see Figure 2.3(b) and (c)). Description of modified Hartmann number (Q) on Bejan number (Be) profile is shown in Figure 2.3(e). The increasing trend of the Be is noticed with an augmentation in Q.

Figure 2.4(a)–(e) displays the trend of temperature profile (θ_1) for radiation constatnt (R_d), temperature ratio constant (θ_w), heat genaration constant (H_g), thermal conjuate constant (Nw), and Dufour number (D_F). Figure 2.4(a) delineates the tendency of θ_1 against R_d. One can note that for larger radiation values, temperature profile (θ_1) shows increasing tendency. For higher amounts of temperature ratio constant $(\theta_w = 0.0,\ 0.2,\ 0.4,\ 0.6)$, increasing trend in liquid temperature is observed (see Figure 2.4(b)). Figure 2.4(c) discloses the effect of heat generation constant (H_g) on θ_1. As we enhance the values of $(H_g = -0.3,\ -1.0,\ 0.0,\ 0.1)$, more heat energy is produced in the system; therefore, θ_1 increases. Figure 2.4(d) indicates the variation of thermal conjugate constant (Nw) on θ_1. An enhance in Nw leads to boosting up the heat transfer coefficient h_f. While Nw depends on the h_f, it transfers more heat from the hot to cold surface of the liquid. Total temperature of the liquid increases with extra heat transfer from the surface to liquid. Figure 2.4(e) outlines the impact of Dufour number (D_F) on θ_1. For higher assessment of $(D_F = 0.0,\ 0.4,\ 0.8,\ 1.2)$, temperature of the liquid rises. An upsurge in heat flux is noted for improved values of D_F. This temperature rise is due to the improvement in concentration gradient.

Impact of Schmidt number (S_C) and Soret number (S_r) on concentration profile (ϕ_1) is shown in Figure 2.5(a) and (b). Figure 2.5(a) elucidates the effect of S_C on ϕ_1. For growing values of $(S_C = 0.0,\ 0.2,\ 0.4,\ 0.6)$, ϕ_1 reduces. Actually, mass diffusivity reduces for higher S_C which concludes that ϕ_1 reduces. Figure 2.5(b) reveals the influence of Soret number (S_r) on ϕ_1. For larger assessment of S_r, concentration and associated boundary layer enhances.

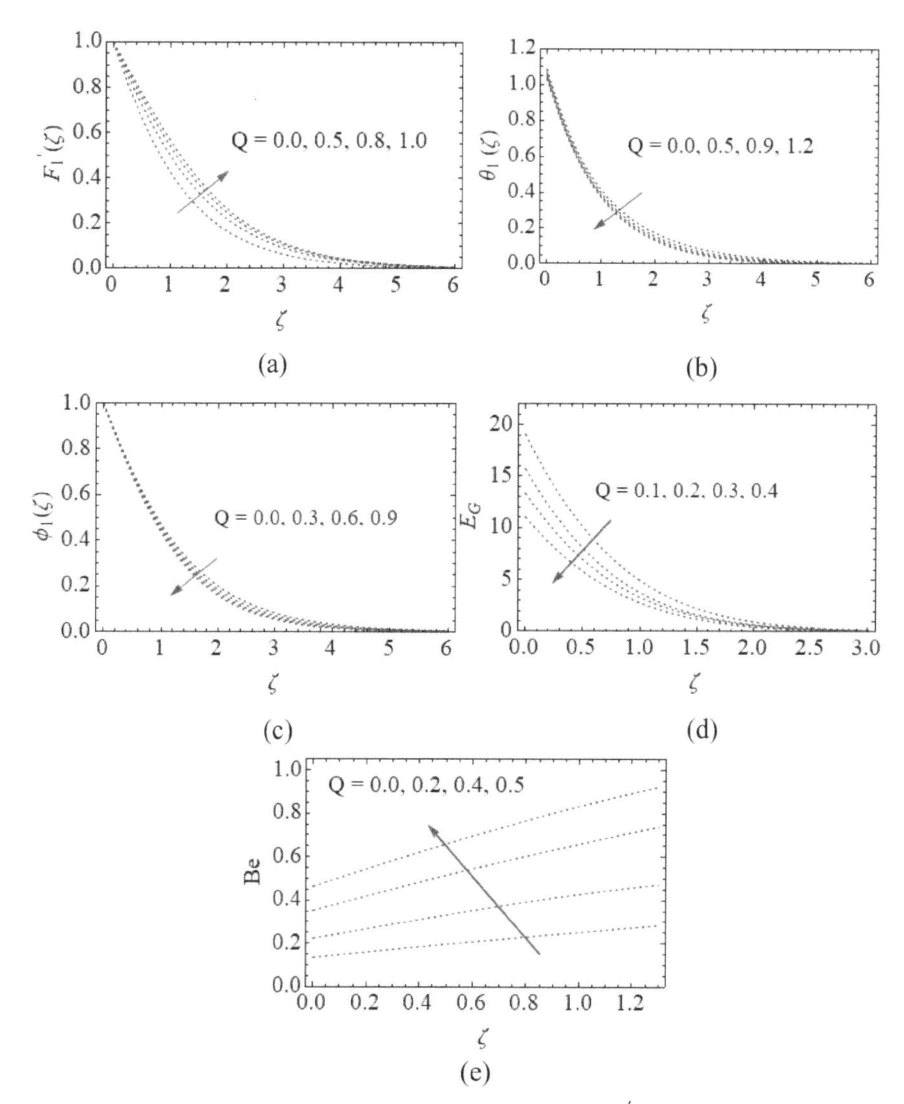

Figure 2.3 Influence of modified Hartmann number (Q) on $F_1^{'}(\zeta)$, $\theta_1(\zeta)$, $\phi_1(\zeta)$, E_G, and Be.

Figure 2.6(a)–(c) is explained to show the tendency of entropy generation (E_G). Figure 2.6(a) sketched to express the impact of Dufour number (D_F) on E_G. We conclude from this figure higher values of D_F initially entropy rate increases and attain the peak value at $\zeta = 3$ and suddenly its fall down.

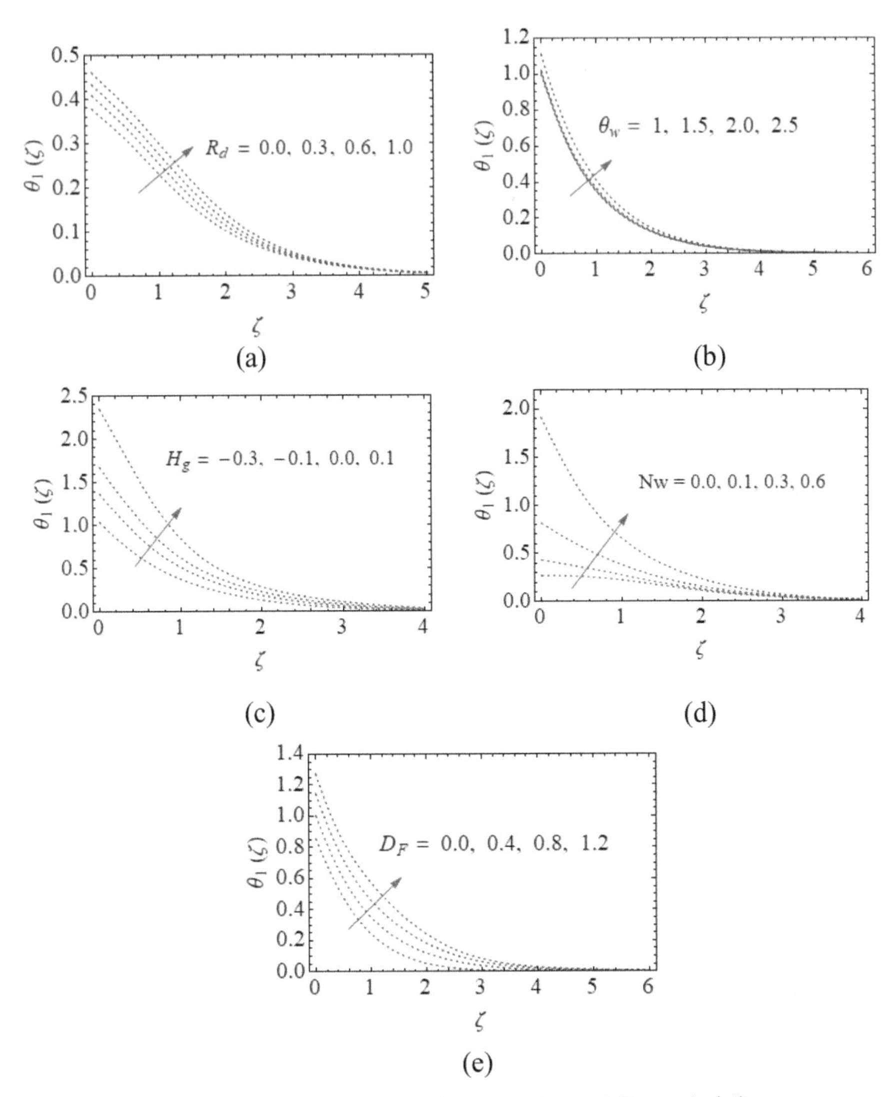

Figure 2.4 Influence of R_d, θ_w, H_g, Nw, and D_F on $\theta_1(\zeta)$.

Again, the entropy rate attains a highest value at $\zeta = 9$ and suddenly falls and remains in the constant form $\zeta = 14$. The same phenomena can be observed for thermal conjugate constant (Nw) and Brinkman number (Br) in entropy profile (see Figure 2.6(b) and (c)).

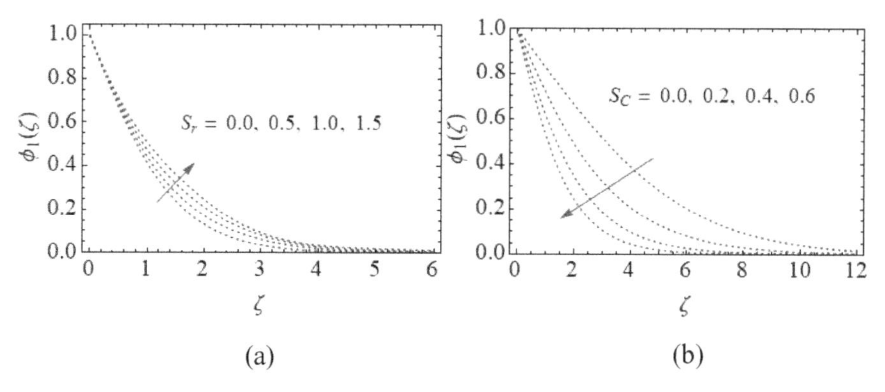

Figure 2.5 Influence of S_r and S_C on $\phi_1(\zeta)$.

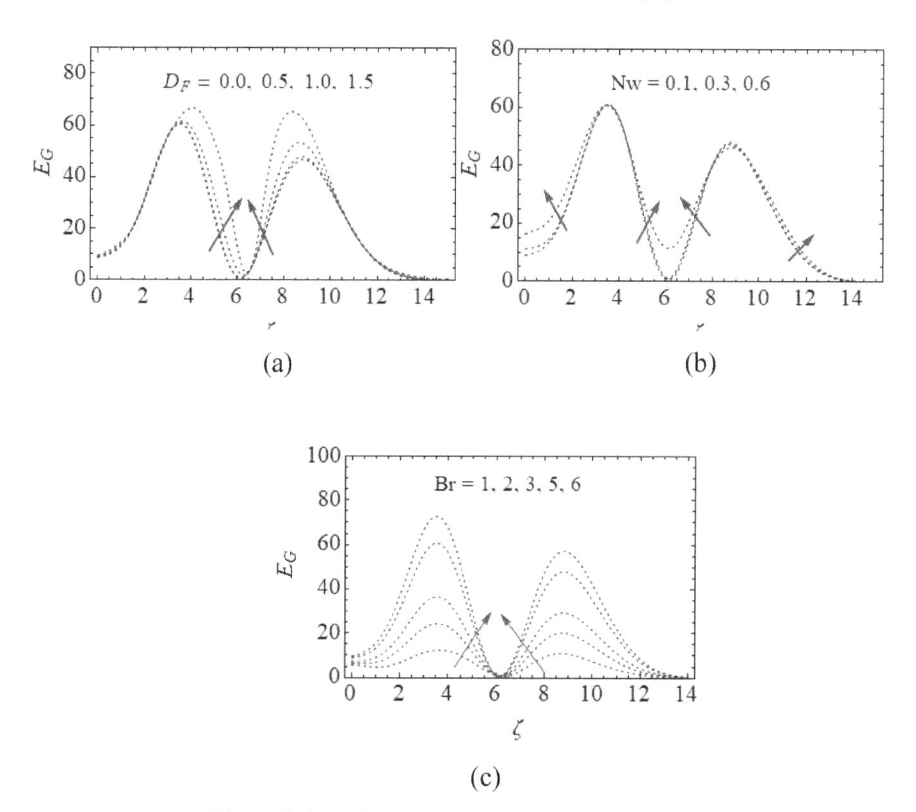

Figure 2.6 Influence of D_F, Nw, and Br on E_G.

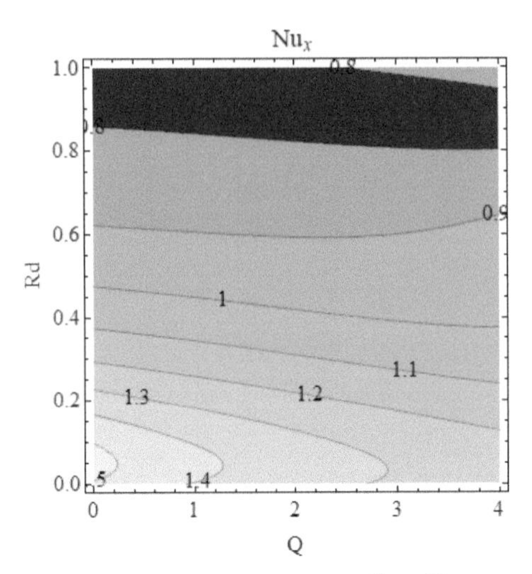

Figure 2.7 Effect of R_d and Q on Nu_x.

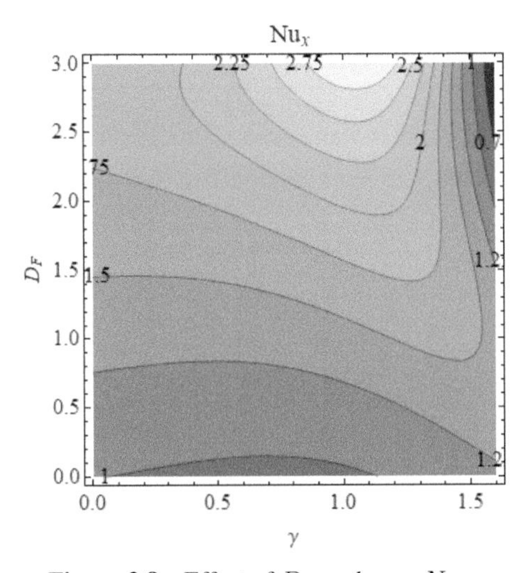

Figure 2.8 Effect of D_F and γ on Nu_x.

Figures 2.7 and 2.8 depict the heat transfer rate (Nu_x) for different combinations of pertinent parameters. From Figure 2.7, heat transfer rate enhances for higher values of modified Hartmann number (Q) and it decays for radiation parameter (R_d). Figure 2.8 elucidates that heat transfer rates reduce for larger values of D_F and H_g. Figures 2.9 and 2.10 display the impact of mass transfer for various combination of parameters. Figure 2.9 explored the

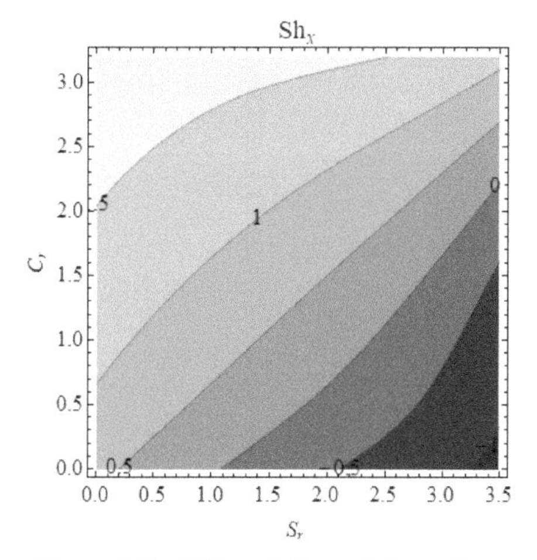

Figure 2.9 Effect of C_r and $S_r$on Sh_x.

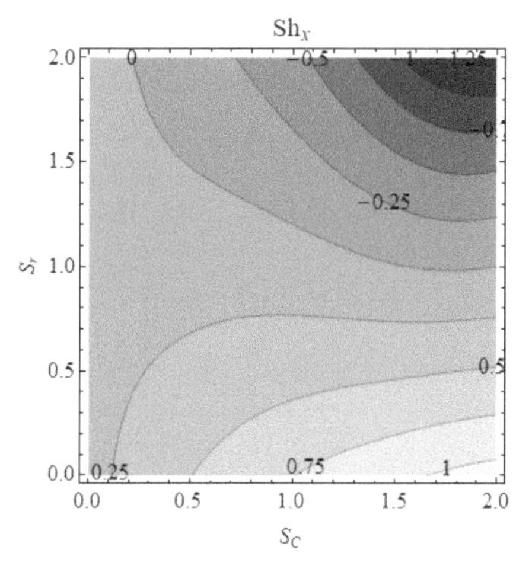

Figure 2.10 Effect of S_r and $S_C$on Sh_x.

combination of S_r and C_r in mass transfer rate (Sh$_\text{x}$). It is noted from this figure that mass transfer rate rises for C_r and it decays for S_r. Figure 2.10 displays the influence of S_r and S_C in Sh$_\text{x}$. From this figure, we conclude that mass transfer rate reduces for the combined parameters S_r and S_C. Table 2.2 indicates the code validation of skin friction ($C_{F_{1x}}$) with formerly available results in [14]. The obtained results found a good agreement.

Table 2.2 Comparison of $Re^{\frac{1}{2}}CF_{1x}$ for different values when $Q = 0$.

α_1	α_2	β_1	Re	Imtiaz [14]	Present
0.0	0.1	0.1	0.1	0.04605	0.04605
0.1			1.	1.06680	1.06680
0.2			2.	1.17470	1.17470
0.1	0.0	0.1	0.1	1.12010	1.12010
	0.1		3.	1.06680	1.06680
	0.2		4.	1.01830	1.01830
0.1	0.1	0.0	5.	1.06290	1.06290
		0.1	6.	1.06680	1.06680
		0.2	7.	1.07030	1.07030
		0.1	0.0	1.06290	1.06290
			0.1	1.06680	1.06680
			0.2	1.07060	1.07060

2.6 Key Results

In the current study, 2D third-grade fluid flows over a Riga plate with cross-diffusion, nonlinear radiation, and Newtonian heating effects are addressed. The key observations are noted below:

- ➢ Higher modified Hartmann number increases the velocity and Bejan number profiles and the opposite tendency noted for concentration, temperature, and entropy profiles.
- ➢ Dufour number acts as a key role in escalating the temperature and entropy profiles.
- ➢ Concentration boundary layer boosts for higher Soret number.
- ➢ Heat transfer rate reduces for Dufour number and thermal relaxation time.

References

[1] M.M. Rashidi, M. Ali, N. Freidoonimehr, F. Nazari. 2013. Parametric analysis and optimization of entropy generation in unsteady MHD flow over a stretching rotating disk using artificial neural network and particle swarm optimization algorithm. Energy. 55:497–510.

[2] A. Malvandi, F. Hedayati, D.D. Ganji. 2013. An analytical study on entropy generation of nanofluids over a flat plate, Alex. Eng. J. 52:277–283.

[3] F. Hedayati, A. Malvandi, D.D. Ganji. 2014. Second-law analysis of fluid flow over an isothermal moving wedge, Alex. Eng. J. 53:1–9.

[4] M.M. Rashidi, S. Abelman, N. Freidoonimehr. 2013. Entropy Generation in Steady MHD Flow Due to a Rotating Porous Disk in a Nanofluid Int. J. Heat Mass Transf. 62:515–525.

[5] K. Loganathan, S. Rajan. 2020. An entropy approach of Williamson nanofluid flow with Joule heating and zero nanoparticle mass flux, J. Therm Anal Calorim. 141:2599–2612.

[6] S. Qayyum, M.I. Khan, T. Hayat, A. Alsaedi, M. Tamoor. 2018. Entropy generation in dissipative flow of Williamson fluid between two rotating disks. Int J Heat Mass Transf. 127:933–42.

[7] K. Loganathan, K. Mohana, M. Mohanraj, P. Sakthivel, S. Rajan. 2020. Impact of third-grade nanofluid flow across a convective surface in the presence of inclined Lorentz force: an approach to entropy optimization, J. Therm Anal Calorim, https://doi.org/10.1007/s10973-020-09751-3.

[8] T. Hayat, M.I. Khan, S. Qayyum, A. Alsaedi. 2018. Entropy generation in flow with silver and copper nanoparticles. Colloids Surf A. 539:335–46.

[9] M.R. Hajmohammadi, S.S. Nourazar. 2014. On the solution of characteristic value problems arising in linear stability analysis; semi analytical approach. Appl Math Comput, 239:126–132.

[10] M.R. Hajmohammadi, S.S. Nourazar. 2014. Conjugate forced convection heat transfer from a heated flat plate of finite thickness and temperature-dependent thermal conductivity. Heat Transf. Eng, 35:863–874.

[11] N. Freidoonimehr, M.M. Rashidi, S. Mahmud. 2015. Unsteady MHD free convective flow past a permeable stretching vertical surface in a nano-fluid, Int. J. Therm. Sci, 87:136–145.

[12] S. Liao, Y.A. Tan. 2007. General approach to obtain series solutions of nonlinear differential. Stud Appl Math, 119:297–354.

[13] K. Loganathan, S. Sivasankaran, M. Bhuvaneshwari, S. Rajan. 2019. Second-order slip, cross-diffusion and chemical reaction effects on magneto-convection of Oldroyd-B liquid using Cattaneo–Christov heat flux with convective heating., J. Therm Anal Calorim, 136:401–409.

[14] M. Imtiaz, A. Alsaedi, A. Shafiq, T. Hayat. 2017. Impact of chemical reaction on third grade fluid flow with Cattaneo–Christov heat flux. J. Mol Liq, 229:501–7.

[15] K. Loganathan, G. Muhiuddin, AM. Alanazi, FS. Alshammari, B M. Alqurashi and S. Rajan. 2020. Entropy Optimization of Third-Grade Nanofluid Slip Flow Embedded in a Porous Sheet With Zero Mass Flux and a Non-Fourier Heat Flux Model. Front. Phys. 8:250.

3

Canonical Representation of Koga's Solution to the Dirac Equation

S. K. Pandey

Department of Mathematics, SPUP, Jodhpur (Raj.), India.
E-mail: skpandey12@gmail.com

Abstract

Geometric algebra provides useful tools and techniques to study image processing, pattern recognition, computer vision, computer graphics and quantum physics etc. However this work deals with the application of geometric algebra to quantum physics. In the geometric algebra treatment of the Dirac equation each even multivector satisfying the Dirac-Hestenes equation has the canonical representation. In this chapter we provide and discuss the canonical representation of Toyoki Koga's solution to the Dirac equation.

Keywords: Geometric algebra, spacetime algebra, Dirac equation, determinism, even multivector, electron, Dirac-Hestenes equation, Koga's theory.

Mathematics Subject Classification 2020: 81R25, 81Q99. 81R99.

3.1 Introduction

Geometric (Clifford) algebras have wide spread applications in computer sciences, engineering, robotics, mathematics and physics etc. [1–6].

Present work is based on the application of geometric algebra to the Dirac equation. This work is an extension of our earlier works published in [7–9].

Toyoki Koga [10–13] has studied the Dirac equation and he has given a deterministic solution to the Dirac equation. As per Koga [10–12] conventional solution to the Dirac equation represents an ensemble of electrons and his solution represents a single electron.

David Hestenes has presented a wide study on quantum mechanics using geometric algebra [14–18]. He has translated the usual Dirac equation in the language of geometric algebra for the first time in 1966 [17] and now this equation is known as the Dirac-Hestenes equation.

In the past we have studied Koga's solution to the Dirac equation using the technique of geometric algebra introduced by Hestenes and obtained very significant results [7–9].

The purpose of this chapter is to provide the canonical representation of the Koga's solution to the Dirac equation. The canonical representation of an even multivector satisfying the Dirac-Hestenes equation is given by $\psi = (\rho\, e^{I\beta})^{\frac{1}{2}}\, R_{rot}$ [14–16]. Here ρ and β are real scalars. As per Hestenes [14–15] the term $(\rho\, e^{I\beta})^{\frac{1}{2}}$ has statistical interpretation but in this chapter we note that his interpretation is meaningful only in the conventional interpretation of the quantum mechanics.

In section two we provide some basic results. It is required to understand this work properly and in section three we provide canonical representation of the Koga's solution to the Dirac equation. In section four we provide result and section five deals with discussion about the result. In the last section we provide conclusion.

3.2 Preliminaries

The Dirac equation for a free electron (with $\hbar = c = 1$) can be put in the following form [6]

$$i\hat{\gamma}^{\mu}\partial_{\mu}\Psi = m\Psi$$

Here $\Psi = \begin{pmatrix} \Psi_1 \\ \Psi_2 \\ \Psi_3 \\ \Psi_4 \end{pmatrix}$ is a four spinor and m is the mass of the electron. The operator ∂_{μ} is given by

$$\partial_{\mu} = \frac{\partial}{\partial t} + \frac{\partial}{\partial x} + \frac{\partial}{\partial y} + \frac{\partial}{\partial z}.$$

The coordinate of a point is given by

$$(t,\ x,\ y,\ z) = (x^0,\ x^1,\ x^2,\ x^3).$$

The Minkowski metric has the following form

$$\eta_{ij} dx^i dx^j = \left(dx^0\right)^2 - \left(dx^1\right)^2 - \left(dx^2\right)^2 - \left(dx^3\right)^2$$

The standard Dirac matrices $\hat{\gamma}^{\mu}$ are given by [6]

$$\hat{\gamma}^0 = \begin{pmatrix} I_2 & 0 \\ 0 & -I_2 \end{pmatrix},\ \hat{\gamma}^k = \begin{pmatrix} 0 & -\hat{\sigma}_k \\ \hat{\sigma}_k & 0 \end{pmatrix},\ k = 1,\ 2,\ 3.$$

The Pauli matrices are:

$$\hat{\sigma}_1 = \begin{pmatrix} 0 & 1 \\ 1 & 0 \end{pmatrix},\ \hat{\sigma}_2 = \begin{pmatrix} 0 & -i \\ i & 0 \end{pmatrix},\ \hat{\sigma}_3 = \begin{pmatrix} 1 & 0 \\ 0 & -1 \end{pmatrix}.$$

The Dirac matrices satisfy the well-known anti-commutation relation given as under.

$$\hat{\gamma}_i\, \hat{\gamma}_j + \hat{\gamma}_j\, \hat{\gamma}_i = 2\eta_{ij}$$

We have $\hat{\gamma}_i = \eta_{ij}\hat{\gamma}^i$ and $\hat{\gamma}^i = \eta^{ij}\hat{\gamma}_i$. Further we take $\eta^{00} = \eta_{00} = 1$ and $\eta^{11} = \eta_{11} = -1$ and so on.

Toyoki Koga [10–13] has obtained a solution to the Dirac equation which can be written in the following form [9]

$$\Psi = a\ \exp(iS)\ \left\{E\hat{\gamma}^0 + m - i\hat{\gamma}^1 R_x - i\hat{\gamma}^2 R_y - i\hat{\gamma}^3 R_z\right\}\ A_j \exp(i\theta_j)$$

Where $a = \frac{e^{(-\kappa\, r)}}{r}$, $\kappa > 0$.

$r = |\vec{r}|$ and $\vec{r} = x\hat{i} + y\hat{j} + z\hat{k}$, $S = -Et$

$$R_x = x\left(\frac{1}{r^2} + \frac{\kappa}{r}\right), R_y = y\left(\frac{1}{r^2} + \frac{\kappa}{r}\right), R_z = z\left(\frac{1}{r^2} + \frac{\kappa}{r}\right).$$

and A_j and θ_j are real scalars ($j = 0,\ 1,\ 2,\ 3$).

The above expression for Koga's solution to the Dirac equation has been obtained by assuming that the electron is at rest and it is centred at the origin of the coordinate axes.

Hestenes has introduced the notion of spacetime algebra which is the geometric algebra of Minkowski spacetime.

The spacetime algebra is generated by $\gamma_0,\ \gamma_1,\ \gamma_2,\ \gamma_3$ satisfying the following relations [5, 6]

$$\gamma_0^2 = 1$$

$$\gamma_0 \cdot \gamma_k = 0$$

$\gamma_k \cdot \gamma_m = -1$ if $k = m$ and $\gamma_k \cdot \gamma_m = 0$ if $k \neq m, \quad (k,\ m = 1,\ 2,\ 3)$.

Spacetime algebra is a sixteen dimensional real algebra and its members are called multivectors. A multivector is defined as the sum of terms in which each term is a real number times a product of γ_i's $(i = 0, 1, 2, 3)$. If in any term the number of γ_i factors is even then the term is known as an even term. An even multivector is defined as the sum of even terms only. The algebra of even multivectors is an eight dimensional real algebra which forms a subalgebra of the spacetime algebra.

The geometric product of two multivectors a and b is given by [5, 6]

$$ab = a.b + a \wedge b$$

The inner product $a.b$ is given by [5, 6]

$$a.b = \frac{1}{2}(ab + ba)$$

The outer product $a \wedge b$ is given by [5, 6]

$$a \wedge b = \frac{1}{2}(ab - ba)$$

It may be noted that the inner product $a.b$ is commutative and the outer product $a \wedge b$ is anti-commutative. The geometric product ab is associative but not commutative.

The geometric algebra version of the usual Dirac equation given above is known as the Dirac-Hestenes equation. The Dirac-Hestenes equation is given by [5, 6]

$$\nabla\psi\, I\sigma_3 = m\psi\gamma_0$$

We have $\nabla = \gamma^{\mu}\frac{\partial}{\partial x^{\mu}} = \gamma^0\frac{\partial}{\partial x^0} + \gamma^1\frac{\partial}{\partial x^1} + \gamma^2\frac{\partial}{\partial x^2} + \gamma^3\frac{\partial}{\partial x^3}$

$$\gamma_0 = \gamma^0, \gamma_k = -\gamma^k, (k = 1,\ 2,\ 3)$$

$$\sigma_k = \gamma_k\gamma_0,$$

$$I = \gamma_0\gamma_1\gamma_2\gamma_3.$$

I is called the pseudoscalar. The relation between four spinor Ψ satisfying the usual Dirac equation and even multivector ψ satisfying the Dirac-Hestenes equation is well established and has widely been discussed in [5, 6].

Now in the next section, we begin with a solution of the Dirac-Hestenes equation which is a geometric algebra version of Koga's solution to the Dirac equation.

3.3 Canonical Representation of ψ

In the language of geometric algebra Koga's solution to the Dirac equation given in section two takes the following form and satisfies the Dirac-Hestenes equation [7, 8].

$$\psi = \vec{R}\varphi\, I\sigma_3 + (E+m)\;\varphi\gamma_0 \tag{3.1}$$

Here $\varphi = ae^{SI\sigma_3}\gamma_0,\;\; a = \frac{e^{(-\kappa\, r)}}{r}$, $\kappa > 0$.

$$r = |\vec{r}|\,,\; \vec{R} = \vec{r}\left(\frac{1}{r^2} + \frac{\kappa}{r}\right).\; S = -Et \text{ and } \vec{r} = x^1\gamma_1 + x^2\gamma_2 + x^3\gamma_3.$$

For the derivation of the above solution given by (3.1) one may refer [7, 9].

In this chapter we study this solution further and we are mainly interested to find its canonical representation. As per geometric algebra description of the Dirac equation each solution to the Dirac-Hestenes has the canonical representation. Hence (3.1) must also have the canonical representation.

In order to provide the canonical representation of (3.1) we have to find real scalars ρ and β and a rotor R_{rot} such that $\psi \;=\; (\rho\, e^{I\beta})^{\frac{1}{2}}\; R_{rot}$ as mentioned in the introduction of this chapter. For this purpose we have to first compute $\psi\tilde{\psi}$.

An even multivector satisfying the Dirac-Hestenes equation has the following property [5, 6]

$$\psi\tilde{\psi} = \rho e^{I\beta} \tag{3.2}$$

Here $\rho = \rho(x)$ and β are real scalars and $\tilde{\psi}$ denotes the reverse of ψ.

Let

$$M = \alpha + a + B + Ib + I\beta$$

be a general multivector of the spacetime algebra. Then the reverse of M is denoted and defined by [6]

$$\tilde{M} = \alpha + a - B - Ib + I\beta$$

Here α and β are scalars, a and b are vectors and B is a bivector.

In our case ψ is an even multivector and therefore a and b must be zero in this case. This follows directly from the definition of an even multivector described in section two.

The computation of $\psi\tilde{\psi}$ is facilitated using above definitions. After a very careful computation we obtain

$$\psi\tilde{\psi} = a^2\left[\{(E+m) + IR_z\}^2 - \left(R_x^2 + R_y^2\right)\right] \tag{3.3}$$

$$\Rightarrow \rho e^{I\beta} = a^2 \left[\{(E+m) + IR_z\}^2 - \left(R_x^2 + R_y^2\right) \right] \tag{3.4}$$

As explained by Hestenes; $\psi\tilde{\psi}$ must contain scalar and pseudo scalar terms only. One can easily verify that (3.4) contains scalar and pseudo scalar terms only. The scalar term is

$$a^2 \left[(E+m)^2 - \left(R_x^2 + R_y^2 + R_z^2\right) \right]$$

and the pseudoscalar term is

$$2a^2 \ (E+m) \ IR_z.$$

The rotor R_{rot} is defined as $R_{rot} = \left(\rho e^{I\beta}\right)^{\frac{1}{2}} \psi$ [5, 6]. Using this we define the rotor R_{rot} as given under.

$$R_{rot} = \frac{\vec{R}\varphi\, I\sigma_3 + (E+m)\ \varphi\gamma_0}{a \left[\{(E+m) + IR_z\}^2 - \left(R_x^2 + R_y^2\right) \right]^{\frac{1}{2}}} \tag{3.5}$$

It is easy to verify that $R_{rot}\, \tilde{R}_{rot} = \tilde{R}_{rot}\, R_{rot} = 1$.

Now we can express ψ as under

$$\psi = \frac{a \left[\{(E+m) + IR_z\}^2 - \left(R_x^2 + R_y^2\right) \right]^{\frac{1}{2}} \left[\vec{R}\varphi\, I\sigma_3 + (E+m)\ \varphi\gamma_0 \right]}{a \left[\{(E+m) + IR_z\}^2 - \left(R_x^2 + R_y^2\right) \right]^{\frac{1}{2}}} \tag{3.6}$$

$\Rightarrow \psi = \left(\rho\, e^{I\beta}\right)^{\frac{1}{2}} \dfrac{\vec{R}\varphi\, I\sigma_3 + (E+m)\ \varphi\gamma_0}{a \left[\{(E+m)+IR_z\}^2 - \left(R_x^2 + R_y^2\right)\right]^{\frac{1}{2}}}$ [Using (3.4)]

$\Rightarrow \psi = \left(\rho\, e^{I\beta}\right)^{\frac{1}{2}} R_{rot}$ [Using (3.5)].

Clearly (3.6) gives canonical representation of ψ.

In the next section we provide discussions and conclusions.

3.4 Result

We note that

$$\psi = \frac{a \left[\{(E+m) + IR_z\}^2 - \left(R_x^2 + R_y^2\right) \right]^{\frac{1}{2}} \left[\vec{R}\varphi\, I\sigma_3 + (E+m)\ \varphi\gamma_0 \right]}{a \left[\{(E+m) + IR_z\}^2 - \left(R_x^2 + R_y^2\right) \right]^{\frac{1}{2}}}$$

is the canonical representation of Koga's solution to the Dirac equation.

In the next section we shall discus about this result.

3.5 Discussions

It should be noted that ρ and β are uniquely determined by (3.4) and in this case rotor R_{rot} is uniquely given by (3.5). Hence the canonical representation of ψ as given by (3.6) is unique.

So far canonical representation of Koga's solution was not given. Here for the first time we have given the canonical representation of Koga's solution to the Dirac equation.

Following the conventional quantum mechanics Hestenes [14–16] has called $\left(\rho\, e^{I\beta}\right)^{\frac{1}{2}}$, which appears in the canonical representation of ψ, as statistical factor. But Koga's theory [10–13] is a deterministic theory of electron as in this theory the position of the electron is given as a function of position and time. Therefore Hestenes's interpretation is not applicable in this case.

3.6 Conclusions

In the canonical representation of Koga's solution to the Dirac equation factor $(\rho\, e^{I\beta})^{\frac{1}{2}}$ does not seem to have any proper physical meaning because Koga's theory is a deterministic theory of the electron. We note that the interpretation given by Hestenes is applicable only in the case of conventional Copenhagen interpretation of quantum mechanics as Koga's theory suggests that probabilistic interpretation of quantum physics is a manifestation of mistaking ensemble solution as a single particle solution.

Acknowledgements

This author is highly thankful to his teachers and friends B. Seeta Ram and A. Pandit for their kind support.

References

[1] J. Vince, Geometric Algebra for Computer Graphics, Springer-Verlag, 2008.

[2] E. Bayro-Corrochano, Geometric Algebra Applications Vol. I: Computer Vision, Graphics and Neurocomputing, Springer International Publishing, 2019.

[3] C. Perwass, et. al., Geometric Algebra with Applications in Engineering, Springer-Verlag Berlin Heidelberg, 2009.
[4] J. Vince, Geometric Algebra: An Algebraic System for Computer Games and Animation, Springer-Verlag, London, 2009.
[5] W. E. Baylis, Clifford Geometric Algebras with Applications to Physics, Mathematics and Engineering, Birkhauser Basel, 1996.
[6] C. Doran, A. Lasenby, Geometric Algebra for Physicists, Cambridge University Press, 2003.
[7] S. K. Pandey, R. S. Chakravarti, The Dirac equation: an approach through geometric algebra, Annales de la Fondation Louis de Broglie, 34 (2), 223–228, 2009.
[8] S. K. Pandey, R. S. Chakravarti, The Dirac equation through geometric algebra: some implications, Annales de la Fondation Louis de Broglie, 36 , 73–77, 2011.
[9] S. K. Pandey, Electronic Spin: Abstract Mathematical or Real Physical Phenomenon, arXiv: 1208.5764 [quan-ph], 2014.
[10] T. Koga, Foundations of Quantum Physics, Woods and Jones, California, 1981.
[11] T. Koga, Inquiries into Foundations of Quantum Physics, Woods and Jones, California, 1983.
[12] T. Koga, A Rational Interpretation of the Dirac Equation for the Electron, Int. J. Theo. Physics, 13 (4), 271–278, 1975.
[13] T. Koga, Representation of Spin in the Interpretation of the Dirac Equation for the Electron, Int. J. Theo. Physics, 12, 205–215, 1975.
[14] D. Hestenes, Mysteries and Insights of Dirac Theory, Annales de la Fondation Louis de Broglie, 28 (3–4), 367–389, 2003.
[15] D. Hestenes, The Zitterbewegung Interpretation of Quantum Physics, Found. Physics, 20 (10), 1213–1232, 1990.
[16] D. Hestenes, Spacetime Physics with Geometric Algebra, Am. J. Phys. 71 (7), 691–714, 2003.
[17] D. Hestenes, Spacetime Algebra, Gordon and Breach, New York, 1966.
[18] D. Hestenes, Spin and Isospin, J. Math. Phys., 8,798, 1967.

4

Bounds of Generalized 'Useful' Information Measure With Application of Jensen's Inequality

Pankaj Prasad Dwivedi and D. K. Sharma

Jaypee University of Engineering and Technology, A.B. Road, Raghogarh, Guna, Madhya Pradesh 473226, India
E-mail: dilipsharmajiet@gmail.com

Abstract

'Useful' information measure is one of the generalizations of Shannon's entropy when utilities are attached to probabilities. The classical state of Jensen's inequality comprises many numbers and weights. The inequality can be expressed quite mostly using either the communication of measure theory or chance. In the probabilistic mounting, the inequality can be far general to its overladen power. In this chapter, first, 'useful' information measure and Jensen's inequality along with utility are defined and explained. A convex function is considered, and then lower and upper bounds for Jensen's inequality along with utility are developed. We also establish bounds on Shannon's information measure with application of bounds obtained for Jensen's inequality.

Keywords: Shannon information, Jensen's inequality, Local bounds, Utility distribution, 'Useful' information inequalities, Convex function.

4.1 Introduction

Information is an integral component of the message or message transmission process. In our daily lives, knowledge was used with or without understanding well before computers found their way into our lives. Any entity that wishes to live in its environment must integrate (independently) food, energy, or, quite generally, environmental knowledge which is essential for it to exist. This begins with basic data types, becomes more important for plants and animals, and eventually culminates in human beings who, due to the invention of a complex language, are the only creatures able to express abstract concepts and ideas. The transfer of knowledge between human beings (i.e., the distribution, through our own senses, of commonly known characters, signs, symbols, noises, and other trends) has evolved from basic reactions to our complex languages. Due to some kind of behavior of single members, the tendency of herds to escape is an instance of such basic reactions, transmitting knowledge from one individual to the entire herd.

Suppose $\triangle_n^+ = \{p = (p_1, p_2, \ldots, p_n);\ p_i \geq 0,\ \sum_i^n p_i = 1\}$ be a lot of all conceivable discrete likelihood appropriations of an arbitrary variable X having utility distribution $U = \{(u_1, u_2, \ldots, u_n);\ u_i > 0\ \forall i\}$ attached to each $P \in \triangle_n^+$ such that $u_i > 0$ is the utility of an occasion having likelihood of event $p_i > 0$.

On the off chance that u_i is the utility of outcome x_i at that point, it is independent of the likelihood of encoding of source symbol x_i, i.e., p_i.

As a consequence, the source of information is given by:

$$\begin{bmatrix} x_1, x_2 \ldots\ldots\ldots x_n \\ p_1, p_2 \ldots\ldots\ldots p_n \\ u_1, u_2 \ldots\ldots\ldots u_n \end{bmatrix}, \ u_i > 0, \quad 0 < p_i \leq 1, \sum_{i=1}^{n} p_i = 1 \tag{4.1}$$

which is a calculation for the normal amount of information 'useful' in the sequence provided by the source (4.1). Belis and Guiasu [2] saw that a source isn't totally indicated by the likelihood distribution P over the source images X without the subjective character of it. Along these lines, it can likewise be accepted that the source letters or images are allocated loads as per their significance or utilities considering the experimenter. Consequently, Belis and Guisau [2] presented the accompanying subjective quantitative proportion of information:

$$H(P;U) = -\sum_{i=1}^{n} u_i p_i log\, p_i. \tag{4.2}$$

It is clear that when utilities are ignored, (4.2) reduces to Shannon's information measure [19] which is given in the following:

$$H(P) = -\sum_{i=1}^{n} p_i log\, p_i. \tag{4.3}$$

Different authors have portrayed the Shannon's entropy by utilizing various hypotheses. Khinch in [13] offered Shannon's expression more accurate by utilizing significant presumptions which were found by Fadeev [7]. Shannon's entropy was additionally characterized by Tverberg [20], Chandy and Mcliod [4], Kendall [10], etc., by considering different sets of postulates.

Inequalities in mathematical analysis and comparison have now become a significant cornerstone and have found applications in a number of settings. An increasing research on this subject offers new evidence and new variations. Also, the improvements, generalizations, numerous approximations, and applications are taken into consideration in. For example, the concept of results and, consequently, the theory of knowledge. In recent years, several new studies on presumptions, extensions, and implementations of inequality have also emerged. In addition, in many different areas of the mathematical, physical, and scientific purposes, analytic as well as geometric inequalities are potentially useful. In probability theory, economic theory, mathematical modeling, information theory, etc., the important function played by Jensen's classical inequality is well known. A number of authors have found, in past years, the prospect of extending such inequality with the grouped convex functions (in the context of the presence of one inflexion point) structure. One presumes that certain conditions of symmetry are confirmed by both the function and the measure under consideration. The Jensen inequality, however, seems to be more general.

Iscan [9] has studied general integral inequalities for quasi-geometrically convex functions via fractional integrals. Kunt and Işcan [14] have gone through fractional Hermite–Hadamard–Fejer type inequalities for GA-convex functions. Association of Jensen inequality for s-convex function was mainly focused by Khan *et al.* [1]. Jensen's type inequalities with generalized majorization inequalities were explored by Khan *et al.* [12]. Simpson's type inequalities for strongly (s, m) (s, m)-convex functions in second sense, and also its applications, was studied by Kermausuor [11]. Fejer-type integral inequalities for geometrically and arithmetically convex functions were developed and applied by Latif *et al.* [15].

On the off chance that $\tilde{p} = \{p_i\}_1^n$, $\sum_1^n p_i = 1$, is a positive weight arrangement, $U = \{u_i\}_1^n$ are the utilities attached to probabilities $\tilde{p} = \{p_i\}_1^n$, and there is a grouping $X = \{x_i\}_1^n, x_i \in [l,\ m], i = 1, 2 \ldots, n$; at that point, the Jensen's inequality along with utility expresses that:

If g is convex on $:= [l,\ m]$, then

$$0 \leq \frac{\sum_{1}^{n} u_i p_i g(x_i)}{\sum_{1}^{n} u_i p_i} - g\left(\frac{\sum_{1}^{n} u_i p_i x_i}{\sum_{1}^{n} u_i p_i}\right). \tag{4.4}$$

When utilities are ignored, the above inequality reduces to inequality given by Simic [18], and the sign of equality holds under the condition that all the elements of X are equal. g holds, linear on $[l,\ m]$ as by Hardy *et al.* [8, p. 70] and Mitrinovic [16].

However, converse of the above statement is also valid. Subsequently, the lower bound zero in (4.4) is of overall disposition; it falls under just upon g and E, and does not rely upon groupings $\tilde{p}$ and X. The structure of this chapter is as follows. In Section 4.2, we develop the lower and upper bounds for Jensen's inequality with utility and show how the improvement in lower bound can be made. In Section 4.3, we establish bounds on 'useful' information measure with application of bounds obtained for Jensen's inequality.

4.2 Lower and Upper Bounds for Jensen's Inequality With Utility

Theorem 4.1. If g is convex on E, then

$$0 \leq \max_{1 \leq \delta < t \leq n} \left[p_\delta g(x_\delta) + p_t g(x_t) - (p_\delta + p_t) g\left(\frac{p_\delta x_\delta + p_t x_t}{p_\delta + p_t}\right)\right]$$

$$\leq \frac{\sum_{1}^{n} u_i p_i g(x_i)}{\sum_{1}^{n} u_i p_i} - g\left(\frac{\sum_{1}^{n} u_i p_i x_i}{\sum_{1}^{n} u_i p_i}\right) \tag{4.5}$$

and this bound is sharp.

Proof: Pick discretionary $x_r, x_s \in \tilde{x},\ 1 \leq r < s \leq n$, with comparing loads $p_r, p_s \in \tilde{p}$. Note that on the off chance$x_r, x_s \in E$, at that point, likewise $\frac{p_r x_r + p_s x_s}{p_r + p_s} \in E$.

By (4.4), we get

$$g\left(\frac{\sum_{1}^{n} u_i p_i x_i}{\sum_{1}^{n} u_i p_i}\right) = g\left(\frac{\sum_{i\neq r,s}^{n} u_i p_i x_i}{\sum_{i\neq r,s}^{n} u_i p_i} + (p_r + p_s)\left(\frac{p_r x_r + p_s x_s}{p_r + p_s}\right)\right)$$

$$\leq \frac{\sum_{i\neq r,s}^{n} u_i p_i g(x_i)}{\sum_{i\neq r,s}^{n} u_i p_i} + (p_r + p_s) g\left(\frac{p_r x_r + p_s x_s}{p_r + p_s}\right).$$

Hence,

$$\frac{\sum_{1}^{n} u_i p_i g(x_i)}{\sum_{1}^{n} u_i p_i} - g\left(\frac{\sum_{1}^{n} u_i p_i x_i}{\sum_{1}^{n} u_i p_i}\right) \geq p_r g(x_r) + p_s g(x_s)$$
$$- (p_r + p_s) g\left(\frac{p_r x_r + p_s x_s}{p_r + p_s}\right). \tag{4.6}$$

When utility is ignored, the above inequality reduces to inequality, given by Simic [18].

Since $x_r, x_s \in \tilde{x}$ are licentious, hence (4.5) holds good.

Note that for $n = 2$, there is the uniformity sign in (4.6). A similar circumstance occurs for $n > 2$ and arbitrary definite $r,\ s$. Simply take $x_i = \frac{p_r x_r + p_s x_s}{p_r + p_s}$ for $i \neq r,\ s$.

In this way, the lower bound in Theorem 4.1 is precipitant.

A worldwide upper bound for a differentiable raised mapping is given by Dragomir in [5].

In the event that g is a differentiable raised mapping on E, at that point, we have

$$0 \leq \frac{\sum_{1}^{n} u_i p_i g(x_i)}{\sum_{1}^{n} u_i p_i} - g\left(\frac{\sum_{1}^{n} u_i p_i x_i}{\sum_{1}^{n} u_i p_i}\right) \leq \frac{1}{4}(m - l)(g^{'}(m) - g^{'}(l))$$
$$:= D_g(l,\ m). \tag{4.7}$$

There isan impressive number of utilizations of this affirmation in information theory (cf., by Budimir *et al.* [3], Dragomir [5], and Dragomir and Goh [6]).

In the continuation, we will give an opposite of Jensen's inequality without a differentiability presumption on g.

Theorem 4.2. For any $\tilde{p}$ and $\tilde{x} \in [l,\ m]$, then

$$\frac{\sum_1^n u_i p_i g(x_i)}{\sum_1^n u_i p_i} - g\left(\frac{\sum_1^n u_i p_i x_i}{\sum_1^n u_i p_i}\right) \leq g(l) + g(m) - 2g\left(\frac{l+m}{2}\right)$$
$$:= S_g(l,\ m). \tag{4.8}$$

In certain instances, the bound $S_g(l,\ m)$ is more efficient than $D_g(l,\ m)$.

Proof: Since $x_i \in [l,\ m]$, there is a grouping $\{\lambda_i\}$, $\lambda_i \in [0, 1]$, such that $x_i = \lambda_i l + (1 - \lambda_i)m$. Consequently,

$$\begin{aligned}
\frac{\sum_1^n u_i p_i g(x_i)}{\sum_1^n u_i p_i} - g\left(\frac{\sum_1^n u_i p_i x_i}{\sum_1^n u_i p_i}\right) &= \frac{\sum_1^n u_i p_i g(\lambda_i l + (1-\lambda_i)m)}{\sum_1^n u_i p_i} \\
&\quad - g\left(\frac{\sum_1^n u_i p_i(\lambda_i l + (1-\lambda_i)m)}{\sum_1^n u_i p_i}\right) \\
&\leq \frac{\sum_1^n u_i p_i(\lambda_i g(l) + (1-\lambda_i)g(m))}{\sum_1^n u_i p_i} \\
&\quad - g\left(l\frac{\sum_1^n u_i p_i \lambda_i}{\sum_1^n u_i p_i} + m\frac{\sum_1^n u_i p_i(1-\lambda_i)}{\sum_1^n u_i p_i}\right) \\
&= g(l)\left(\frac{\sum_1^n u_i p_i \lambda_i}{\sum_1^n u_i p_i}\right) + g(m)\left(1 - \frac{\sum_1^n u_i p_i \lambda_i}{\sum_1^n u_i p_i}\right)
\end{aligned}$$

$$-g\left(l\left(\frac{\sum\limits_1^n u_i p_i \lambda_i}{\sum\limits_1^n u_i p_i}\right)+m\left(1-\frac{\sum\limits_1^n u_i p_i \lambda_i}{\sum\limits_1^n u_i p_i}\right)\right).$$

Defining $\frac{\sum_1^n u_i p_i \lambda_i}{\sum_1^n u_i p_i} := p; 1 - \frac{\sum_1^n u_i p_i \lambda_i}{\sum_1^n u_i p_i} := q$, we have note that $0 \le p,\ q \le 1; p+q = 1$ and

$$\frac{\sum\limits_1^n u_i p_i g(x_i)}{\sum\limits_1^n u_i p_i} - g\left(\frac{\sum\limits_1^n u_i p_i x_i}{\sum\limits_1^n u_i p_i}\right) \le pg(l) + qg(m) - g(pl + qm). \quad (4.9)$$

But,

$$\begin{aligned}pg(l)+qg(m)-g(pl+qm) &= g(l)+g(m)-(qg(l)+pg(m))-g(pl+qm)\\ &\le g(l)+g(m)-(g(ql+pm)+g(pl+qm))\\ &\le g(l)+g(m)-2g\left(\frac{1}{2}(ql+pm)+\frac{1}{2}(pl+qm)\right)\\ &= g(l)+g(m)-2g\left(\frac{l+m}{2}\right).\end{aligned}$$

Therefore, by (4.9), the result follows.

For example, taking $g(x) = -x^s,\ 0 < s < 1; g(x) = x^s,\ s > 1; E \subset R^+$, we have that

$$S_g(l,\ m) \le D_g(l,\ m),$$

for each $s \in (0, 1) \cup (1, 2) \cup (3,\ +\infty)$.

4.3 Bounds on 'Useful' Information Measure With Application of Bounds Obtained for Jensen's Inequality

Definition 4.1. If the likelihood distribution F is given by

$P(X = i) = p_i, p_i > 0, i = 1, 2, \ldots, r; \sum_1^r p_i = 1$, then the 'useful' information measure is given as

$$K(P, U) := -\frac{\sum\limits_1^r u_i p_i \log p_i}{\sum\limits_1^r u_i p_i}.$$

It can be noted that when utilities are ignored, then the above measure reduces (4.3).

Proposition 4.1. Define $\delta := \min_{1\leq i\leq r}(p_i)$; $t := \max_{1\leq i\leq r}(p_i)$.

Then

$$(0 \leq) c(\delta,\, t) := \delta \log\left(\frac{2\delta}{\delta+t}\right) + t\log\left(\frac{2\delta}{\delta+t}\right)$$
$$\leq \log r - K(P,U) \leq \log\left(\frac{(\delta+t)^2}{4\delta t}\right) := N(\delta,\, t). \quad (4.10)$$

Proof of Proposition 4.1. Applying Theorems 4.1 and 4.2 with $g(x) = -\log x,\ x_i = 1/p_i\ ,\ i = 1, 2, \ldots,\ r; p_r = \delta = l,\ \ p_s = t = m$, after certain estimations, the ideal affirmation follows.

Remark 4.1. When applied to $(x) = -\log x,\ x_i = 1/p_i\ ,\ i = 1, 2, \ldots,\ r$, Dragomir's result (4.7) implies

$$0 \leq \log r - K(P,U) \leq \frac{(t-\delta)^2}{4\delta t} := D(\delta,\, t). \quad (4.11)$$

Since $\log(1+x) < x,\ x > 0$, putting $x = (t-\delta)^2/4t\delta$, it follows that $N(\delta,\, t) < D(\delta,\, t)$, i.e., the assessment (4.10) is better than (4.11).

Remark 4.2. In 1997, Dragomir and Goh [6], by a fairly confused contention, the creators, acquired the accompanying outcome: In the event that $\frac{t}{\delta} \leq \varphi_1(\varepsilon) := 1 + \varepsilon + \sqrt{\varepsilon(2+\varepsilon)},\ \ \varepsilon > 0$, at that point

$$0 \leq \log r - K(P,U) \leq \varepsilon.$$

By basic analytics in right-hand side of (4.10), we get

Proposition 4.2. In the event that, for some $\varepsilon > 0$,

$$t/\delta \leq \varphi_2(\varepsilon) := 1 + 2(e^{\varepsilon} - 1) + 2\sqrt{e^{2\varepsilon} - e^{\varepsilon}}.$$

Then

$$0 \leq \log r - K(P,U) \leq \varepsilon.$$

Since $\varphi_2(\varepsilon) \gg \varphi_1(\varepsilon)$, by the above recommendation, the time period is amplified.

Similarly, using (4.5) and (4.6), few latest and extra exact limits for the restrictive entropy and internecine data can be set up by Dragomir and Goh (cf., [6]). This is left to the reader.

By using (4.10), we can obtain the desired result.

Proposition 4.3. Under the documentation of Proposition 4.1, we have

$$c(\delta,\ t) \leq \ \log r - K(P,U) \leq \ \min\left\{N(\delta,\ t),\ rc(\delta,\ t)\right\}. \tag{4.12}$$

Proof of Proposition 4.3. Applying Theorem 4.3 with $g(x) = x\ log\ x$ and putting $x_i = p_i,\ ,\ i = 1, 2, \ldots,\ r$, we get

$$\frac{1}{r}\left(\delta \log \delta + t \log t - (\delta + t) \log \left(\frac{\delta + t}{2}\right)\right) \leq \frac{1}{r}\left(\sum_{1}^{r} p_i \log p_i\right) - \frac{1}{r} \log \frac{1}{r}$$

$$\leq \delta \log \delta + t \log t - (\delta + t) \log \left(\frac{\delta + t}{2}\right),$$

which is equal to

$$c(\delta,\ t) \leq \ \log r - K(P,U) \leq rc(\delta,\ t).$$

Combining this with (4.10), we acquire the affirmation from Proposition 4.3.

For the utilization of the above outcomes in the information theory, we will give new limits for 'useful' information measure $K(P;U)$.

Remark 4.3. Using elementary mean values, the bounds c, N can be expressed as

$$c(\delta,\ t) = 2A(\delta,\ t) \log \frac{J(\delta, t)}{A(\delta, t)};\ N(\delta,\ t) = 2 \log \frac{A(\delta, t)}{G(\delta, t)},$$

where

$$A(l,\ m) := \frac{l + m}{2}, G(l,\ m) := \sqrt{lm}, J(l,\ m) := (l^l m^m)^{\frac{1}{l+m}},$$

two positive numbers, l and m, are represented by their arithmetic, geometric, and Gini means.

The subsequent power series interpretations can be deduced using a result from [17]:

$$c(\delta,\ t) = (\delta + t) \sum_{1}^{\infty} \frac{1}{2n(2n-1)} \left(\frac{t - \delta}{t + \delta}\right)^{2n};$$

$$N(\delta,\ t) = \sum_{1}^{\infty} \frac{1}{n} \left(\frac{t - \delta}{t + \delta}\right)^{2n}.$$

Theorem 4.3. If g is convex on E and $\delta := \min_{1 \le i \le n} x_i; t := \max_{1 \le i \le n} x_i; [\delta, t] \subseteq E$, at that point,

$$\frac{1}{n}\left(g(\delta) + g(t) - 2g\left(\frac{\delta + t}{2}\right)\right) \le \frac{1}{n}\sum_{1}^{n} g(x_i) - g\left(\frac{\sum_{1}^{n} x_i}{n}\right)$$
$$\le g(\delta) + g(t) - 2g\left(\frac{\delta + t}{2}\right). \quad (4.13)$$

Proof: As a result of the Theorem 4.2 affirmations, we acquire this outcome. The connection (4.6) is legitimate for arbitrary $p_r,\ x_r; p_s,\ x_s$, putting $p_i = 1/n,\ ,\ i = 1, 2, \ldots,\ n, ; x_r = \delta = l, x_s = t = m$, outcome follows from (4.5) and (4.6).

4.4 Results and Discussions

In this chapter, we developed the lower and upper bounds for Jensen's inequality with utility and showed how the improvement in lower bound can be made. We have also developed new lower and upper bounds on 'useful' information measure with application of bounds obtained for Jensen's inequality. Results have been given in Sections 4.2 and 4.3.

4.5 Conclusion

New upper bounds for Jensen's inequality with utility have been formulated as a direct application of the results obtained by Simic [18] from new generalized inequalities for *n*-convex functions obtained. We also showed how the progress in lower bound can be made. We have also established bounds on 'useful' information measure with application of bounds obtained for Jensen's inequality. We can also construct several functionals from the inequalities introduced in Theorem 4.1 and developed new theorems for convex functions.

References

[1] M. Adil Khan, M. Hanif, Z.A. Khan, K. Ahmad, Y.-M. Chu, 'Association of Jensen inequality for s-convex function,' J. Inequal. Appl. 2019, Article ID 162, 2019.

[2] M. Belis, and S. Guiasu, 'A quantitative-qualitative measure of information in Cybernetics System', IEEE Trans. Inform. Theory, IT 14, pp. 593–594, 1968.

[3] I. Budimir, S.S. Dragomir, J. Pecaric, 'Further reverse results for Jensen's discrete inequality and applications in information theory', J. Inequal. Pure Appl. Math. 2 (1) Art. 5, 2001.

[4] T.W. Chandy, and J.B. Mcliod, 'On a functional equation', Proc. Edinburgh Maths 43, pp. 7–8, 1960.

[5] S.S. Dragomir, 'A converse result for Jensen's discrete inequality via Gruss inequality and applications in information theory', An. Univ. Oradea. Fasc. Mat. 7, pp. 178–189, 1999–2000.

[6] S.S. Dragomir, C.J. Goh, 'Some bounds on entropy measures in Information Theory', Appl. Math. Lett. 10 (3), pp. 23–28, 1997.

[7] D.K. Fadeev, 'On the concept of entropies of finite probabilistic scheme', (Russian) Uspchi Math. Nauk 11, pp. 227–231, 1956.

[8] G.H. Hardy, J.E. Littlewood, G. Polya, 'Inequalities, Camb'., Univ. Press, Cambridge 1978.

[9] I. Iscan, 'New general integral inequalities for quasi-geometrically convex functions via fractional integrals', J. Inequal. Appl. 2013 (491), 1, 2013.

[10] D.G. Kendall, 'Functional equations in information theory', Z.WahrsVerw.Geb 2, pp. 225–229, 1964.

[11] S. Kermausuor, 'Simpson's type inequalities for strongly (s, m) (s, m)-convex functions in second sense and applications', Open J. Math. Sci. 3(1), pp. 74–83, 2019.

[12] J. Khan, M.A. Khan, J. Pecaric, 'On Jensen's type inequalities via generalized majorization inequalities,' Filomat 32(16), pp. 5719–5733, 2018.

[13] A.I. Khinchin, 'Mathematical Foundations of Information Theory', Dover Publications, New York, 1957.

[14] M. Kunt, and I. Işcan, 'Fractional Hermite–Hadamard–Fejer type inequalities for GA-convex functions', Turk. J. Ineq., 2(1), pp. 1–20, 2018.

[15] M.A. Latif, S.S. Dragomir, E. Momoniat, 'Some Fejer type integral inequalities for geometrically–arithmetically-convex functions with applications', Filomat 32(6), pp. 2193–2206, 2018.

[16] D.S. Mitrinovic, 'Analytic Inequalities', Springer, New York, 1970.

[17] J. Sandor, I. Rasa, 'Inequalities for certain means in two arguments', Nieuw Arch. Wiscunde 15, pp. 51–55, 1997.

[18] S. Simic, 'Jensen's inequality and new entropy bounds', Applied Mathematics Letters, 22, pp. 1262–1265, 2009.
[19] C.E. Shannon, 'A mathematical theory of communication', Bell System Technical Journal 27, pp. 379–423 (Part I), 623–656 (Part II), 1948.
[20] H. Tverberg, 'A new derivation of the information function', Math. Scand, 6, pp. 297–298, 1958.

5

Scalar Potential in Planar Filippov Systems

Tomáš Hanus and Drahoslava Janovská

University of Chemistry and Technology, Prague, Czech Republic.
E-mail: tuark@seznam.cz; janovskd@vscht.cz

Abstract

In this chapter, our goal is to apply the tools of qualitative analysis of smooth dynamical systems to piecewise smooth dynamical systems, namely to Filippov systems. The scalar functions (scalar potential, Hamiltonian, Lyapunov function) can describe the behavior of Filippov systems. Using qualitative methods, it is possible to examine phase portraits, trajectories, equilibria, stability, bifurcations, and other local and global properties of Filippov systems.

Keywords: Scalar potential, Filippov system, piecewise smooth, dynamical system, conservative vector field, trajectory, fall line.

5.1 Introduction

By the most common way, a dynamical system with lumped parameters is given by a system of ordinary differential equations (ODEs) in a domain X. If the differential equations are explicit of the first order, then their right-hand sides represent the vector field in the state space X.

The smooth dynamical system has the continuous vector field. Smooth dynamical systems are very well described in theory and contained in a broad spectrum of literature such as [1, 10, 14, 15, 21, 26]. The piecewise smooth dynamical system has the piecewise continuous vector field. Piecewise smooth dynamical systems and their theory can be found, for example, in [3, 5, 11, 12, 19, 20].

Piecewise smooth dynamical systems arise in a large number of applications. In the mechanics of moving bodies, jumps of speed, acceleration, and force can occur due to impact or friction. In electronics and information technology: pulse control, feedback control, optimum regulation, two position regulation, etc. In biology and ecology: anthropogenic effects on populations, sudden environmental changes, biosystems with critical transitions, etc. Piecewise smooth mathematical models are suitable for these dynamics; see [2, 4, 6, 7, 8, 9, 22, 23, 24, 25, 27].

We have discovered some applications of piecewise smooth dynamical systems in chemistry: a continuously stirred tank reactor with an outfall, a system of chemical reactions with a zero-order reaction, a system of reactions with a reversible enzyme denaturation, a synthesis with temporary lack of reactants, etc. These phenomena can be mathematically modeled as Filippov systems; see [13, 16, 17, 18].

5.2 Simple Planar Filippov System

Among all dynamical systems, we will focus on piecewise smooth dynamical systems. Among all piecewise smooth dynamical systems, we will focus on Filippov systems. For the first approach to the Filippov systems, we will accept some simplifications. We will limit the topological dimension of the state space, we will reduce the quantities of the unions and the boundaries, and we will modify their properties.

We will introduce a simple planar Filippov system under the following assumptions.

1) Let the state space be the plane, $S = \mathbb{R}^2$.
2) Let the state space S contain two sets, called unions, $S_1 \subset S$ and $S_2 \subset S$.

In Filippov systems, the unions S_1 and S_2 are open in S. They are disjoint, i.e., $S_1 \cap S_2 = \emptyset$, and their closure covers S, i.e., $\overline{S_1 \cup S_2} \supset S$. The universal union S_0 is $S_0 = S_1 \cup S_2$. The universal boundary Σ_0 is $\Sigma_0 = \partial S_0$. The boundary Σ_1 of the union S_1 is $\Sigma_1 = \partial S_1$. The boundary Σ_2 of the union S_2 is $\Sigma_2 = \partial S_2$. The boundary Σ_{12} between the unions S_1 and S_2 is $\Sigma_{12} = \partial S_1 \cap \partial S_2$.

3) Let the unions S_1 and S_2 be regular open in S,
   i.e., $S_1 = \mathrm{int}(\overline{S_1})$ and $S_2 = \mathrm{int}(\overline{S_2})$.

In the regular open set, some parts of the boundary are excluded. In the regular open union S_1, the boundary Σ_1 lies only on the surface of the union S_1.

In the regular open union S_2, the boundary Σ_2 lies only on the surface of the union S_2. As a consequence, the boundary Σ_{12} is the only boundary in the state space S, i.e., $\Sigma_0 = \Sigma_{12}$.

4) Let the boundary Σ_{12} be the simple boundary.

Definition 5.1. *The simple boundary in the plane is the curve in $\mathbb{R}^2$, whose connected components are formed by:*

1) closed curves,
2) open curves,

and which:

a) has only regular points,
b) is nowhere dense set in $\mathbb{R}^2$,
c) is closed set in $\mathbb{R}^2$.

A closed curve is a set of points topologically equivalent to the circle. An open curve is a set of points topologically equivalent to the straight line. If there are always two or more points of intersection between the open curve and the circle with any radius centered at any point of this open curve, then this open curve ends at infinity. The regular point is the point of the curve, which has the open neighborhood, on which this curve is diffeomorphic with the straight line open segment. The singular point is any point of the curve, which is not regular.

Remark 5.1. *Typical singular points are: isolated point (acnode), end point (terminal), tangent point (tacnode), crossing point (crunode), cusp (spinode), spike (apex), multiple point (node), etc.*

If the given curve is the simple boundary, then it is the differentiable curve and it consists of connected components, called branches. If the given curve is the simple boundary, then it is the boundary and also the complement of the set, which is open, dense in $\mathbb{R}^2$.

Definition 5.2. *Let the set S be given. Let its subsets S_1, S_2, Σ_{12} be given, $S_1 \subset S$, $S_2 \subset S$, $\Sigma_{12} \subset S$. Let these sets have the following properties:*

a) $S = \mathbb{R}^2$,
b) $S_1 \cup S_2 \cup \Sigma_{12} = \mathbb{R}^2$,
c) $S_1 \cap S_2 = \emptyset$,
d) $S_1 \cap \Sigma_{12} = \emptyset$,

e) $S_2 \cap \Sigma_{12} = \emptyset$,
f) S_1 *is regular open in* $\mathbb{R}^2$,
g) S_2 *is regular open in* $\mathbb{R}^2$,
h) Σ_{12} *is the simple boundary in the plane.*

The set S *equipped with the sets* S_1, S_2, Σ_{12} *is called the simple planar Filippov state space* S.

The set S_1, called union, consists of connected components $\triangle S_1$, called regions. The set S_2, called union, consists of connected components $\triangle S_2$, called regions. The set Σ_{12}, called boundary, consists of connected components $\triangle\Sigma_{12}$, called branches; see Figure 5.1.

In the simple planar Filippov system, there are given two continuous vector fields, $\mathbf{F}_1 : S_1 \to \mathbb{R}^2$ and $\mathbf{F}_2 : S_2 \to \mathbb{R}^2$, which can be continuously extended to $\overline{S_1}$ and $\overline{S_2}$,

$$\mathbf{F}_1(\mathbf{x}) = \lim_{\substack{\boldsymbol{x} \in S_1 \\ \boldsymbol{x} \to \mathbf{x}}} \mathbf{F}_1(\boldsymbol{x}), \quad \mathbf{x} \in \partial S_1\,, \qquad \mathbf{F}_2(\mathbf{x}) = \lim_{\substack{\boldsymbol{x} \in S_2 \\ \boldsymbol{x} \to \mathbf{x}}} \mathbf{F}_2(\boldsymbol{x})\,, \;\; \mathbf{x} \in \partial S_2\,.$$

In the simple planar Filippov system, there is also given a scalar function $h : S \to \mathbb{R}$ such that it decides whether a particular point of S belongs to S_1

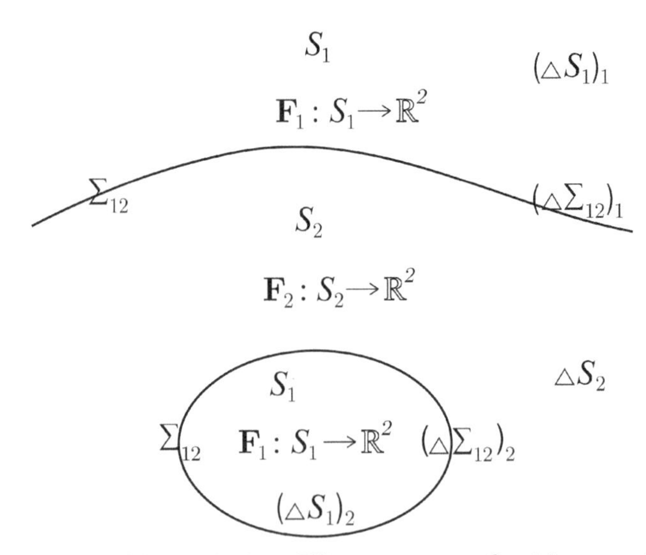

Figure 5.1 Example of the simple planar Filippov state space S and the vector fields $\mathbf{F}_1, \mathbf{F}_2$.

or S_2 or Σ_{12} :

$$\begin{array}{lcl} \mathbf{x} \in S_1 & \Leftrightarrow & h(\mathbf{x}) > 0\,, \\ \mathbf{x} \in \Sigma_{12} & \Leftrightarrow & h(\mathbf{x}) = 0\,, \qquad \mathbf{x} \in S\,. \\ \mathbf{x} \in S_2 & \Leftrightarrow & h(\mathbf{x}) < 0\,, \end{array}$$

The function h gives: the union S_1 as the preimage of positives, i.e., $S_1 = h^{-1}(\mathbb{R}^+)$, the boundary Σ_{12} as the zero level set, i.e., $\Sigma_{12} = h^{-1}(0)$, and the union S_2 as the preimage of negatives, i.e., $S_2 = h^{-1}(\mathbb{R}^-)$.

Remark 5.2. *Let an arbitrary function $h : S \to \mathbb{R}$ be chosen. The conditions: h is smooth, $\nabla h(\mathbf{x}) \neq \mathbf{0}$ for all $\mathbf{x} \in h^{-1}(0)$ are sufficient, but not necessary, conditions for the function h to give the sets $h^{-1}(\mathbb{R}^+)$, $h^{-1}(0)$, $h^{-1}(\mathbb{R}^-)$ in accordance with Definition 5.2.*

We denote by $\mathbf{\Sigma}^N(\mathbf{x})$ a non-zero normal vector and by $\mathbf{\Sigma}^T(\mathbf{x})$ a non-zero tangent vector to Σ_{12} at point $\mathbf{x}$. We continuously extend $\mathbf{F}_1$ to $\overline{S}_1$ and $\mathbf{F}_2$ to $\overline{S}_2$. The vector fields $\mathbf{F}_1$ and $\mathbf{F}_2$ on the boundary Σ_{12} imply two vector fields. The vector field by the arithmetic mean $\mathbf{G}_{12}^{AM} : \Sigma_{12} \to \mathbb{R}^2$ is

$$\mathbf{G}_{12}^{AM}(\mathbf{x}) = \frac{1}{2}\left(\mathbf{F}_1(\mathbf{x}) + \mathbf{F}_2(\mathbf{x})\right)\,.$$

The vector field by the Filippov method $\mathbf{G}_{12}^{FM} : \mathbf{\Sigma}_{12} \backslash \Sigma_{12}^{PD} \to \mathbb{R}^2$ is

$$\mathbf{G}_{12}^{FM}(\mathbf{x}) = \frac{\mathbf{\Sigma}^N(\mathbf{x}) \cdot \mathbf{F}_2(\mathbf{x})}{\mathbf{\Sigma}^N(\mathbf{x}) \cdot (\mathbf{F}_2(\mathbf{x}) - \mathbf{F}_1(\mathbf{x}))}\mathbf{F}_1(\mathbf{x}) - \frac{\mathbf{\Sigma}^N(\mathbf{x}) \cdot \mathbf{F}_1(\mathbf{x})}{\mathbf{\Sigma}^N(\mathbf{x}) \cdot (\mathbf{F}_2(\mathbf{x}) - \mathbf{F}_1(\mathbf{x}))}\mathbf{F}_2(\mathbf{x}).$$

At the points with the parallel difference $\mathbf{x} \in \Sigma_{12}^{PD}$, the difference of the vectors $\mathbf{F}_1$, $\mathbf{F}_2$ is parallel to Σ_{12}, i.e., $(\mathbf{F}_2 - \mathbf{F}_1) \parallel \mathbf{\Sigma}^T$. Thus, $\mathbf{\Sigma}^N \cdot (\mathbf{F}_2 - \mathbf{F}_1) = 0$ and $\mathbf{G}_{12}^{FM}$ is undefined. At the points with the double tangency $\mathbf{x} \in \Sigma_{12}^{DT}$, both vectors $\mathbf{F}_1$, $\mathbf{F}_2$ are tangent to Σ_{12}, i.e., $\mathbf{F}_1 \parallel \mathbf{F}_2 \parallel \mathbf{\Sigma}^T$. Thus, $\mathbf{\Sigma}^N \cdot (\mathbf{F}_2 - \mathbf{F}_1) = 0$ and $\mathbf{G}_{12}^{FM}$ is undefined. At the tangent points $\mathbf{x} \in \Sigma_{12}^{TP}$, at least one of the vectors $\mathbf{F}_1$, $\mathbf{F}_2$ is tangent to Σ_{12}, i.e., $\mathbf{F}_1 \parallel \mathbf{\Sigma}^T$ or $\mathbf{F}_2 \parallel \mathbf{\Sigma}^T$.

The vector fields $\mathbf{F}_1$ and $\mathbf{F}_2$ on the boundary Σ_{12} imply four scalar functions $\kappa, \tau, \sigma, \chi : \Sigma_{12} \to \mathbb{R}$:

$$\begin{aligned} \kappa(\mathbf{x}) &= \left(\mathbf{\Sigma}^N(\mathbf{x}) \cdot \mathbf{F}_2(\mathbf{x})\right) - \left(\mathbf{\Sigma}^N(\mathbf{x}) \cdot \mathbf{F}_1(\mathbf{x})\right)\,, \\ \tau(\mathbf{x}) &= \left(\mathbf{\Sigma}^N(\mathbf{x}) \cdot \mathbf{F}_2(\mathbf{x})\right)^2 + \left(\mathbf{\Sigma}^N(\mathbf{x}) \cdot \mathbf{F}_1(\mathbf{x})\right)^2\,, \\ \sigma(\mathbf{x}) &= \left(\mathbf{\Sigma}^N(\mathbf{x}) \cdot \mathbf{F}_2(\mathbf{x})\right)\left(\mathbf{\Sigma}^N(\mathbf{x}) \cdot \mathbf{F}_1(\mathbf{x})\right)\,, \\ \chi(\mathbf{x}) &= \operatorname{sgn}\left(\mathbf{\Sigma}^N(\mathbf{x}) \cdot \mathbf{F}_2(\mathbf{x})\right) - \operatorname{sgn}\left(\mathbf{\Sigma}^N(\mathbf{x}) \cdot \mathbf{F}_1(\mathbf{x})\right)\,. \end{aligned}$$

The functions κ, τ, σ determine whether a particular point of Σ_{12} is

the point with the parallel difference or not:

$$\begin{array}{lcl} \mathbf{x} \notin \Sigma_{12}^{PD} & \Leftrightarrow & \kappa(\mathbf{x}) \neq 0\,, \\ \mathbf{x} \in \Sigma_{12}^{PD} & \Leftrightarrow & \kappa(\mathbf{x}) = 0\,, \end{array} \qquad \mathbf{x} \in \Sigma_{12}\,,$$

the point with the double tangency or not:

$$\begin{array}{lcl} \mathbf{x} \notin \Sigma_{12}^{DT} & \Leftrightarrow & \tau(\mathbf{x}) \neq 0\,, \\ \mathbf{x} \in \Sigma_{12}^{DT} & \Leftrightarrow & \tau(\mathbf{x}) = 0\,, \end{array} \qquad \mathbf{x} \in \Sigma_{12}\,,$$

the tangent point or not:

$$\begin{array}{lcl} \mathbf{x} \notin \Sigma_{12}^{TP} & \Leftrightarrow & \sigma(\mathbf{x}) \neq 0\,, \\ \mathbf{x} \in \Sigma_{12}^{TP} & \Leftrightarrow & \sigma(\mathbf{x}) = 0\,, \end{array} \qquad \mathbf{x} \in \Sigma_{12}\,.$$

The function σ decides whether a particular point of Σ_{12} belongs to the crossing set Σ_{12}^C or to the sliding set Σ_{12}^S :

$$\begin{array}{lcl} \mathbf{x} \in \Sigma_{12}^{C} & \Leftrightarrow & \sigma(\mathbf{x}) > 0\,, \\ \mathbf{x} \in \Sigma_{12}^{S} & \Leftrightarrow & \sigma(\mathbf{x}) \leq 0\,, \end{array} \qquad \mathbf{x} \in \Sigma_{12}\,.$$

The function σ gives: the crossing set Σ_{12}^C as the preimage of positives, i.e., $\Sigma_{12}^C = \sigma^{-1}(\mathbb{R}^+)$, and the sliding set Σ_{12}^S as the preimage of non-positives, i.e., $\Sigma_{12}^S = \sigma^{-1}(\mathbb{R}^- \cup \{0\})$.

At the points of the crossing set Σ_{12}^C, both vectors $\mathbf{F}_1$, $\mathbf{F}_2$ point to the same side of Σ_{12}; see Figure 5.2 on the left. We define the vector field $\mathbf{G}_{12} : \Sigma_{12}^C \to \mathbb{R}^2$ by the arithmetic mean,

$$\mathbf{G}_{12}(\mathbf{x}) = \mathbf{G}_{12}^{AM}(\mathbf{x})\,, \quad \mathbf{x} \in \Sigma_{12}^C\,.$$

At the points of the sliding set Σ_{12}^S, the vectors $\mathbf{F}_1, \mathbf{F}_2$ are at other positions to Σ_{12}; see Figure 5.2 on the right. We define the vector field $\mathbf{G}_{12} : \Sigma_{12}^S \to \mathbb{R}^2$ by the Filippov method, at the points without the parallel difference,

$$\mathbf{G}_{12}(\mathbf{x}) = \mathbf{G}_{12}^{FM}(\mathbf{x})\,, \quad \mathbf{x} \in \Sigma_{12}^S \setminus \Sigma_{12}^{PD}\,,$$

or by the arithmetic mean, at the points with the parallel difference,

$$\mathbf{G}_{12}(\mathbf{x}) = \mathbf{G}_{12}^{AM}(\mathbf{x})\,, \quad \mathbf{x} \in \Sigma_{12}^{PD}\,.$$

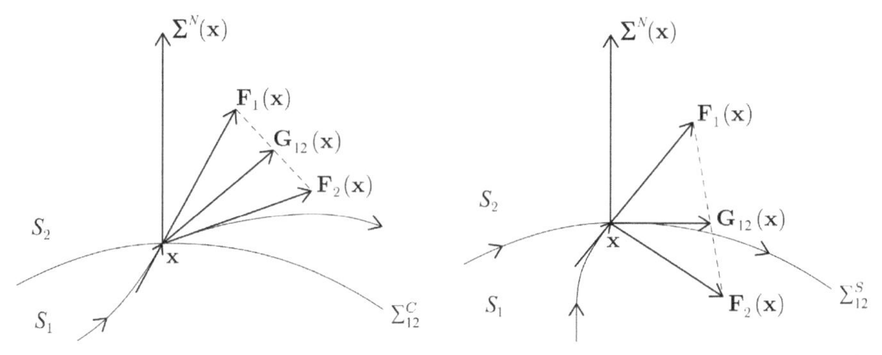

Figure 5.2 On the left: Example of the vectors and the trajectories near the crossing set Σ_{12}^C. On the right: Example of the vectors and the trajectories near the sliding set Σ_{12}^S.

On the whole boundary Σ_{12}, the vector field $\mathbf{G}_{12} : \Sigma_{12} \to \mathbb{R}^2$ is

$$\mathbf{G}_{12}(\mathbf{x}) = \begin{cases} \mathbf{G}_{12}^{AM}(\mathbf{x}), & \mathbf{x} \in \Sigma_{12}^C \cup \Sigma_{12}^{PD}, \\ \mathbf{G}_{12}^{FM}(\mathbf{x}), & \mathbf{x} \in \Sigma_{12}^S \setminus \Sigma_{12}^{PD}, \end{cases} \quad \mathbf{x} \in \Sigma_{12}. \tag{5.1}$$

The function χ determines whether a particular point of Σ_{12} is equipped with the vector $\mathbf{G}_{12}^{AM}$ or with the vector $\mathbf{G}_{12}^{FM}$:

$$\begin{array}{lcl} \mathbf{x} \in \Sigma_{12}^C \cup \Sigma_{12}^{PD} & \Leftrightarrow & \chi(\mathbf{x}) = 0, \\ \mathbf{x} \in \Sigma_{12}^S \setminus \Sigma_{12}^{PD} & \Leftrightarrow & \chi(\mathbf{x}) \neq 0, \end{array} \quad \mathbf{x} \in \Sigma_{12}.$$

Definition 5.3. *Let the simple planar Filippov state space S be given. Let the vector field $\mathbf{F}_1 : S_1 \to \mathbb{R}^2$ be continuous on S_1 and continuously extendable to $\overline{S_1}$. Let the vector field $\mathbf{F}_2 : S_2 \to \mathbb{R}^2$ be continuous on S_2 and continuously extendable to $\overline{S_2}$. These vector fields imply the vector field $\mathbf{G}_{12} : \Sigma_{12} \to \mathbb{R}^2$,*

$$\mathbf{G}_{12}(\mathbf{x}) = \begin{cases} \frac{1}{2}(\mathbf{F}_1(\mathbf{x}) + \mathbf{F}_2(\mathbf{x})), \\ \qquad \operatorname{sgn}(\mathbf{\Sigma}^N(\mathbf{x}) \cdot \mathbf{F}_2(\mathbf{x})) - \operatorname{sgn}(\mathbf{\Sigma}^N(\mathbf{x}) \cdot \mathbf{F}_1(\mathbf{x})) = 0, \ \mathbf{x} \in \Sigma_{12}, \\ \\ \dfrac{\mathbf{\Sigma}^N(\mathbf{x}) \cdot \mathbf{F}_2(\mathbf{x})}{\mathbf{\Sigma}^N(\mathbf{x}) \cdot (\mathbf{F}_2(\mathbf{x}) - \mathbf{F}_1(\mathbf{x}))} \mathbf{F}_1(\mathbf{x}) - \dfrac{\mathbf{\Sigma}^N(\mathbf{x}) \cdot \mathbf{F}_1(\mathbf{x})}{\mathbf{\Sigma}^N(\mathbf{x}) \cdot (\mathbf{F}_2(\mathbf{x}) - \mathbf{F}_1(\mathbf{x}))} \mathbf{F}_2(\mathbf{x}), \\ \\ \qquad \operatorname{sgn}(\mathbf{\Sigma}^N(\mathbf{x}) \cdot \mathbf{F}_2(\mathbf{x})) - \operatorname{sgn}(\mathbf{\Sigma}^N(\mathbf{x}) \cdot \mathbf{F}_1(\mathbf{x})) \neq 0, \ \mathbf{x} \in \Sigma_{12}, \end{cases}$$

where $\mathbf{\Sigma}^N(\mathbf{x})$ *is the non-zero normal vector to* Σ_{12} *at the point* $\mathbf{x}$.
The vector field $\mathbf{F} : S \to \mathbb{R}^2$,

$$\mathbf{F}(\mathbf{x}) = \begin{cases} \mathbf{F}_1(\mathbf{x})\,, & \mathbf{x} \in S_1\,, \\ \mathbf{F}_2(\mathbf{x})\,, & \mathbf{x} \in S_2\,, \\ \mathbf{G}_{12}(\mathbf{x})\,, & \mathbf{x} \in \Sigma_{12}\,, \end{cases} \qquad \mathbf{x} = [x, y] \in S\,, \tag{5.2}$$

is called the simple planar Filippov vector field.

The sum of equations (5.1) and (5.2) is a notation of a simple planar Filippov vector field $\mathbf{F} : S \to \mathbb{R}^2$,

$$\mathbf{F}(\mathbf{x}) = \begin{cases} \mathbf{F}_1(\mathbf{x})\,, & \mathbf{x} \in S_1\,, \\ \mathbf{F}_2(\mathbf{x})\,, & \mathbf{x} \in S_2\,, \\ \mathbf{G}_{12}^{AM}(\mathbf{x})\,, & \mathbf{x} \in \Sigma_{12}^{C} \cup \Sigma_{12}^{PD}\,, \\ \mathbf{G}_{12}^{FM}(\mathbf{x})\,, & \mathbf{x} \in \Sigma_{12}^{S} \setminus \Sigma_{12}^{PD}\,, \end{cases} \qquad \mathbf{x} \in S\,.$$

Definition 5.4. *The simple planar Filippov system is the system of ordinary differential equations with the independent variable t as the time*

$$\frac{\mathrm{d}\mathbf{x}}{\mathrm{d}t} = \mathbf{F}(\mathbf{x})\,, \quad \mathbf{x} = [x, y] \in S\,, \tag{5.3}$$

where $\mathbf{F}$ *is the simple planar Filippov vector field* (5.2).

Remark 5.3. *A simple planar Filippov vector field is given by two vector fields* $\mathbf{F}_1$, $\mathbf{F}_2$ *and by one scalar function* h. *A simple planar Filippov system is given by a simple planar Filippov vector field.*

Remark 5.4. *It holds that* $\Sigma_{12}^{S} \cap \Sigma_{12}^{PD} = \Sigma_{12}^{DT}$. *On the sliding set, the parallel difference is also the double tangency. A notation of a simple planar Filippov vector field* $\mathbf{F} : S \to \mathbb{R}^2$ *is*

$$\mathbf{F}(\mathbf{x}) = \begin{cases} \mathbf{F}_1(\mathbf{x})\,, & \mathbf{x} \in S_1\,, \\ \mathbf{F}_2(\mathbf{x})\,, & \mathbf{x} \in S_2\,, \\ \mathbf{G}_{12}^{AM}(\mathbf{x})\,, & \mathbf{x} \in \Sigma_{12}^{C} \cup \Sigma_{12}^{DT}\,, \\ \mathbf{G}_{12}^{FM}(\mathbf{x})\,, & \mathbf{x} \in \Sigma_{12}^{S} \setminus \Sigma_{12}^{DT}\,, \end{cases} \qquad \mathbf{x} \in S\,.$$

Remark 5.5. *In some cases, there are no points with the double tangency. In some cases, the points with the double tangency are discussed separately. For* $\Sigma_{12}^{DT} = \emptyset$, *a notation of a simple planar Filippov vector field* $\mathbf{F} : S \to \mathbb{R}^2$ *is*

$$\mathbf{F}(\mathbf{x}) = \begin{cases} \mathbf{F}_1(\mathbf{x})\,, & \mathbf{x} \in S_1\,, \\ \mathbf{F}_2(\mathbf{x})\,, & \mathbf{x} \in S_2\,, \\ \mathbf{G}_{12}^{AM}(\mathbf{x})\,, & \mathbf{x} \in \Sigma_{12}^{C}\,, \\ \mathbf{G}_{12}^{FM}(\mathbf{x})\,, & \mathbf{x} \in \Sigma_{12}^{S}\,, \end{cases} \qquad \mathbf{x} \in S\,.$$

5.3 Curve Integral in a Simple Planar Filippov System

Let S be a nonempty, open, connected subset of $\mathbb{R}^2$. Let the continuous vector field $\mathbf{F} : S \to \mathbb{R}^2$ be given. Let the path of integration be the smooth curve $\mathcal{I} \subset S$,

$$\mathcal{I} = \Phi(I)\,,\ \Phi : I \to \mathbb{R}^2\,,\ a, b \in \mathbb{R}\,,\ I = \langle a, b\rangle\,,\ \Phi(a) = A\,,\ \Phi(b) = B\,.$$

The function Φ is the smooth parameterization of the curve $\mathcal{I}$. The curve integral of the vector field $\mathbf{F}$ along the path $\mathcal{I}$ from the point A to the point B is

$$\int_{\mathcal{I}} \mathbf{F} \cdot \mathrm{d}\Phi.$$

Now, let the path of integration be the piecewise differentiable curve $\mathcal{K}$, which is the oriented sum of the differentiable curves $\mathcal{K}_1, \ldots, \mathcal{K}_r$, $r \in \mathbb{N}$,

$$\mathcal{K} = \mathcal{K}_1 \dotplus \cdots \dotplus \mathcal{K}_r\,.$$

The curves $\mathcal{K}_i$ are also the smooth curves. They have the smooth parameterizations Φ_i. They are oriented in accordance with Φ_i.

$$\mathcal{K}_i = \Phi_i(I_i)\,,\quad \Phi_i : I_i \to \mathbb{R}^2\,,\quad a_i, b_i \in \mathbb{R}\,,\quad I_i = \langle a_i, b_i\rangle\,,$$

$$\begin{array}{rcccll} & & A & = A_1 & = \Phi_1(a_1)\,, & \\ \Phi_i(b_i) & = & B_i & = A_{i+1} & = \Phi_{i+1}(a_{i+1})\,, & i = 1, \ldots, r-1\,, \\ \Phi_r(b_r) & = & B_r & = B\,. & & \end{array}$$

The curve integral of the vector field $\mathbf{F}$ along the path $\mathcal{K}$ from the point A to the point B is the sum

$$\int_{\mathcal{K}} \mathbf{F} \cdot \mathrm{d}\Phi = \sum_{i=1}^{r} \int_{\mathcal{K}_i} \mathbf{F} \cdot \mathrm{d}\Phi_i\,.$$

Remark 5.6. *The path of integration can be generalized. The starting point A can be at infinity, i.e., "$\|A\| \to \infty$", and the ending point B can be at infinity, i.e., "$\|B\| \to \infty$". The path $\mathcal{K}$ can consist of infinitely many parts $\mathcal{K}_i$, i.e., "$r \to \infty$".*

The nonempty intersection of the simple boundary Σ_{12} and the piecewise differentiable curve $\mathcal{K}$ has connected components, which consist either of one point (single point) or of uncountably many points (segment of curve).

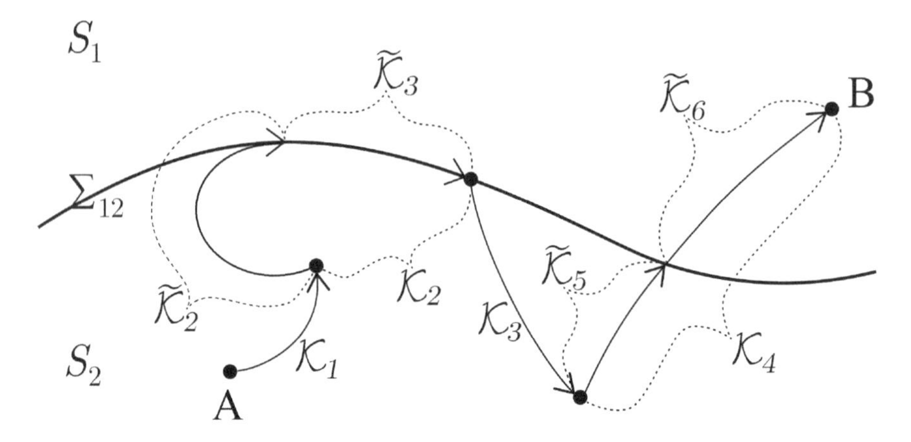

Figure 5.3 The path of integration $\mathcal{K}$ inserted into the simple planar Filippov state space S.

Let the simple planar Filippov state space S be given. Let the simple planar Filippov vector field $\mathbf{F} : S \to \mathbb{R}^2$ be given. Let the path of integration $\mathcal{K} = \mathcal{K}_1 \dotplus \cdots \dotplus \mathcal{K}_r$ be inserted into the state space S; see Figure 5.3. If $\mathcal{K}$ intersects Σ_{12} in the single points, then these points cut $\mathcal{K}$ into new parts $\widetilde{\mathcal{K}}_i$. If the segments of $\mathcal{K}$ lie completely on Σ_{12}, then they also become newly formed parts $\widetilde{\mathcal{K}}_i$. This creates the new division of the path $\mathcal{K}$ into the parts $\widetilde{\mathcal{K}}_1, \ldots, \widetilde{\mathcal{K}}_s$, $s \in \mathbb{N}$,

$$\mathcal{K} = \widetilde{\mathcal{K}}_1 \dotplus \cdots \dotplus \widetilde{\mathcal{K}}_s\,, \quad r \leq s\,.$$

Each part $\widetilde{\mathcal{K}}_i$ is the differentiable curve. The simple planar Filippov vector field $\mathbf{F}$ is continuous on every $\widetilde{\mathcal{K}}_i$, except for some end points A_i, B_i.

Definition 5.5. *The curve integral of the simple planar Filippov vector field* $\mathbf{F} : S \to \mathbb{R}^2$ *in* (5.2) *along the path* $\mathcal{K}$ *from the point* A *to the point* B *is the sum*

$$\int_{\mathcal{K}} \mathbf{F} \cdot \mathrm{d}\Phi = \sum_{i=1}^{s} \int_{\widetilde{\mathcal{K}}_i} \mathbf{F} \cdot \mathrm{d}\Phi_i\,, \tag{5.4}$$

where $\mathcal{K} = \widetilde{\mathcal{K}}_1 \dotplus \cdots \dotplus \widetilde{\mathcal{K}}_s$. *Each part* $\widetilde{\mathcal{K}}_i$ *is the differentiable curve, is the smooth curve, has the smooth parameterization* Φ_i, *and is oriented in accordance with it.*

5.4 Scalar Potential in a Simple Planar Filippov System

Let us recall some basic statements about continuous vector fields, smooth dynamical systems, and topographic surfaces.

Let S be an open region in $\mathbb{R}^2$. A continuous vector field $\mathbf{F} : S \to \mathbb{R}^2$ is said to be conservative on S if and only if there exists a function $U : S \to \mathbb{R}$ such that $\mathbf{F}(\mathbf{x}) = \nabla U(\mathbf{x})$ for all $\mathbf{x} = [x, y] \in S$, where the gradient $\nabla U(\mathbf{x}) = [\frac{\partial U}{\partial x}, \frac{\partial U}{\partial y}]$. The function U is called the scalar potential of the continuous vector field $\mathbf{F}$.

Let S be an open, simply connected subset of $\mathbb{R}^2$. The continuous vector field $\mathbf{F} : S \to \mathbb{R}^2$, $\mathbf{F}(x, y) = [F_x(x, y), F_y(x, y)]$ is conservative on S if and only if $\frac{\partial F_x}{\partial y}(x, y) = \frac{\partial F_y}{\partial x}(x, y)$ for all $[x, y] \in S$.

Let S be an open region in $\mathbb{R}^2$. The continuous vector field $\mathbf{F} : S \to \mathbb{R}^2$ is conservative on S if and only if the curve integral of this vector field is path independent.

Let S be an open region in $\mathbb{R}^2$. Let the continuous vector field $\mathbf{F} : S \to \mathbb{R}^2$ be conservative with the scalar potential $U : S \to \mathbb{R}$. Then the fall lines of the scalar potential U are identical with the trajectories of the smooth dynamical system $\frac{\mathrm{d}\mathbf{x}}{\mathrm{d}t} = \mathbf{F}(\mathbf{x})$, $\mathbf{x} = [x, y] \in S$.

The horizontal trace of the topographic surface U is the line connecting the points of the surface in the direction of its zero ascent. The contour line is the orthogonal projection of the horizontal trace into the horizontal plane S. The line of greatest slope of the topographic surface U is the line connecting the points of the surface in the direction of its steepest ascent. The fall line is the orthogonal projection of the line of greatest slope into the horizontal plane S. The contour lines and the fall lines form a system of orthogonal trajectories in the horizontal plane S.

Now, we propose new definitions and theorems about piecewise continuous vector fields and piecewise smooth dynamical systems.

Definition 5.6. *Let the simple planar Filippov state space S be given. A simple planar Filippov vector field $\mathbf{F} : S \to \mathbb{R}^2$ in (5.2) is said to be conservative on S if and only if there exists a function $U : S \to \mathbb{R}$, which fulfills the following conditions:*

$$\begin{aligned} &a) \quad \nabla U(\mathbf{x}) = \mathbf{F}_1(\mathbf{x}) && \forall \mathbf{x} = [x, y] \in S_1\,, \\ &b) \quad \nabla U(\mathbf{x}) = \mathbf{F}_2(\mathbf{x}) && \forall \mathbf{x} = [x, y] \in S_2\,, \\ &c) \quad U(\mathbf{x}) \ \textit{is continuous} && \forall \mathbf{x} = [x, y] \in S\,, \end{aligned}$$

where the gradient $\nabla U(\mathbf{x}) = [\frac{\partial U}{\partial x}, \frac{\partial U}{\partial y}]$. Then the function U is called the scalar potential of the simple planar Filippov vector field $\mathbf{F}$.

Theorem 5.1. *Let the simple planar Filippov state space S be given. The simple planar Filippov vector field $\mathbf{F} : S \to \mathbb{R}^2$ in (5.2) is conservative on S if and only if the following conditions are fulfilled:*

a) The vector field $\mathbf{F}_1$ is conservative on every connected component $\triangle S_1$.

b) The vector field $\mathbf{F}_2$ is conservative on every connected component $\triangle S_2$.

c) The orthogonal projections of $\mathbf{F}_1(\mathbf{x})$ and $\mathbf{F}_2(\mathbf{x})$ into the tangent line to Σ_{12} are identical, $\mathrm{proj}_{\mathbf{\Sigma}^T}\mathbf{F}_1(\mathbf{x}) = \mathrm{proj}_{\mathbf{\Sigma}^T}\mathbf{F}_2(\mathbf{x})$, at all points $\mathbf{x} = [x, y] \in \Sigma_{12}$.

Theorem 5.2. *Let the simple planar Filippov state space S be given. The simple planar Filippov vector field $\mathbf{F} : S \to \mathbb{R}^2$ in (5.2) is conservative on S if and only if the curve integral of this vector field is path independent.*

Theorem 5.3. *Let the simple planar Filippov state space S be given. Let the simple planar Filippov vector field $\mathbf{F} : S \to \mathbb{R}^2$ in (5.2) be conservative with the scalar potential $U : S \to \mathbb{R}$. Then the fall lines of the scalar potential U are identical with the trajectories of the simple planar Filippov system $\frac{\mathrm{dx}}{\mathrm{dt}} = \mathbf{F}(\mathbf{x})$ in (5.3).*

5.5 Scalar Potential on a Closed Region

Definition 5.7. *Let M be an open region in $\mathbb{R}^2$ and let its closure $\overline{M}$ be a closed region in $\mathbb{R}^2$. Let there be given a continuous vector field $\mathbf{V} : M \to \mathbb{R}^2$, which can be continuously extended onto a vector field $\mathbf{V} : \overline{M} \to \mathbb{R}^2$.*

The continuous vector field $\mathbf{V} : \overline{M} \to \mathbb{R}^2$ is said to be conservative on $\overline{M}$ if and only if the continuous vector field $\mathbf{V} : M \to \mathbb{R}^2$ is conservative on M with the scalar potential $U : M \to \mathbb{R}$, which can be continuously extended onto the scalar potential $U : \overline{M} \to \mathbb{R}$.

The open region M and the closed region $\overline{M}$ are illustrated in Figure 5.4.

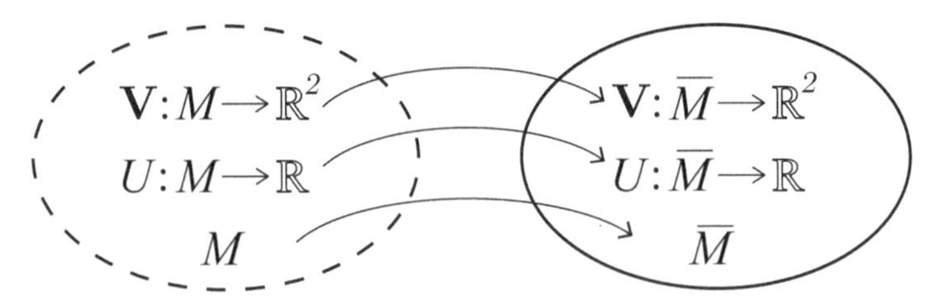

Figure 5.4 The closure of the open region M, and the continuous extension of the vector field $\mathbf{V}$ and of the scalar potential U to the closed region $\overline{M}$.

Lemma 5.1. *Let the set $\overline{M}$ be the closed region in $\mathbb{R}^2$. The continuous vector field $\mathbf{V} : \overline{M} \to \mathbb{R}^2$ is conservative if and only if the curve integral of this vector field is path independent.*

Proof of Lemma 5.1:

i) We assume that the vector field $\mathbf{V} : \overline{M} \to \mathbb{R}^2$ is conservative. Then also $\mathbf{V} : M \to \mathbb{R}^2$ is conservative. We choose an arbitrary closed curve $\mathcal{L} \subset \overline{M}$. We take a sequence of closed curves $\{\mathcal{L}_1, \mathcal{L}_2, \mathcal{L}_3, \dots\}$, $\mathcal{L}_i \subset M$ that converges to $\mathcal{L}$; see Figure 5.5. The curve integral along $\mathcal{L}_i$ is equal to zero for each i. The vector field $\mathbf{V} : \overline{M} \to \mathbb{R}^2$ is continuous; so the limit of the sequence of the curve integrals along $\mathcal{L}_i$ is equal to the curve integral along $\mathcal{L}$. Thus, the curve integral along $\mathcal{L}$ is zero.

In the case i), the curve integral of the vector field $\mathbf{V} : \overline{M} \to \mathbb{R}^2$ is path independent.

ii) Conversely: We assume that the vector field $\mathbf{V} : \overline{M} \to \mathbb{R}^2$ is not conservative. Then two situations may arise:

ii)1) The vector field $\mathbf{V} : M \to \mathbb{R}^2$ is not conservative.

ii)2) The potential $U : M \to \mathbb{R}$ can not be continuously extended onto the potential $U : \overline{M} \to \mathbb{R}$. We choose a point A in the set M. We choose the point B on the boundary of the set $\overline{M}$ such that $U : M \to \mathbb{R}$ can not be

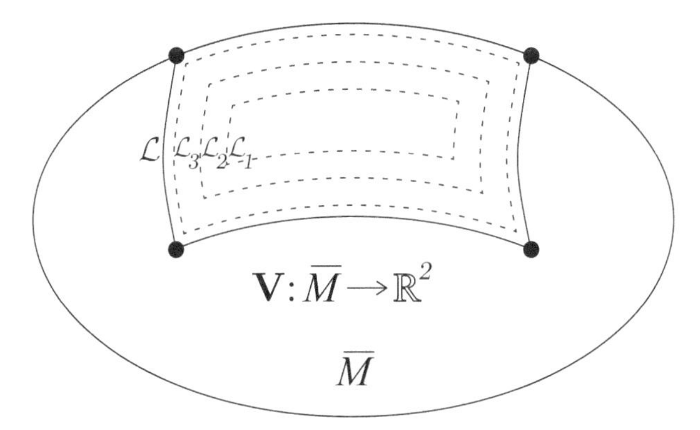

Figure 5.5 The sequence of the closed curves $\mathcal{L}_i$ in the open region M that converges to the closed curve $\mathcal{L}$ in the closed region $\overline{M}$.

continuously extended to the point B. We choose two oriented curves $\widetilde{\mathcal{K}} \in \overline{M}$ and $\widehat{\mathcal{K}} \in \overline{M}$. They start at the same point A, and they end at the same point B, but they follow different paths in order to reach different values of the limits

$$\lim_{\substack{\boldsymbol{x} \in \widetilde{\mathcal{K}} \\ \boldsymbol{x} \to B}} U(\boldsymbol{x}) = \widetilde{U}(B) \neq \widehat{U}(B) = \lim_{\substack{\boldsymbol{x} \in \widehat{\mathcal{K}} \\ \boldsymbol{x} \to B}} U(\boldsymbol{x}) .$$

The curve integrals along the curves $\widetilde{\mathcal{K}}$ and $\widehat{\mathcal{K}}$ give

$$\begin{aligned} \widetilde{U}(B) &= U(A) + \textstyle\int_{\widetilde{\mathcal{K}}} \mathbf{V} \cdot \mathrm{d}\mathbf{r} , \\ \widehat{U}(B) &= U(A) + \textstyle\int_{\widehat{\mathcal{K}}} \mathbf{V} \cdot \mathrm{d}\mathbf{r} . \end{aligned}$$

Because $\widetilde{U}(B) \neq \widehat{U}(B)$, the curve integrals along $\widetilde{\mathcal{K}}$ and $\widehat{\mathcal{K}}$ have different values.

In the cases ii)1), and ii)2), the curve integral of the vector field $\mathbf{V} : \overline{M} \to \mathbb{R}^2$ is path dependent.

□

5.6 Properties of Orthogonal Projections

Lemma 5.2. *Let Σ_{12} be a differentiable curve in $\mathbb{R}^2$. Let $\mathbf{x}$ be a point of Σ_{12}. Let $\mathbf{\Sigma}^T(\mathbf{x})$ be a non-zero tangent vector to Σ_{12} at $\mathbf{x}$. Let the vectors $\mathbf{F}_1(\mathbf{x})$, $\mathbf{F}_2(\mathbf{x})$ have orthogonal projections $\mathrm{proj}_{\mathbf{\Sigma}^T}\mathbf{F}_1(\mathbf{x})$, $\mathrm{proj}_{\mathbf{\Sigma}^T}\mathbf{F}_2(\mathbf{x})$ into the tangent line to Σ_{12} at $\mathbf{x}$.*
The orthogonal projections are identical, $\mathrm{proj}_{\mathbf{\Sigma}^T}\mathbf{F}_1(\mathbf{x}) = \mathrm{proj}_{\mathbf{\Sigma}^T}\mathbf{F}_2(\mathbf{x})$, if and only if the dot products are equal, $\mathbf{F}_1(\mathbf{x}) \cdot \mathbf{\Sigma}^T(\mathbf{x}) = \mathbf{F}_2(\mathbf{x}) \cdot \mathbf{\Sigma}^T(\mathbf{x})$.

Proof of Lemma 5.2:

$$\begin{aligned} \mathrm{proj}_{\mathbf{\Sigma}^T}\mathbf{F}_1 = \mathrm{proj}_{\mathbf{\Sigma}^T}\mathbf{F}_2 \Leftrightarrow \frac{\mathbf{F}_1 \cdot \mathbf{\Sigma}^T}{\mathbf{\Sigma}^T \cdot \mathbf{\Sigma}^T}\mathbf{\Sigma}^T = \frac{\mathbf{F}_2 \cdot \mathbf{\Sigma}^T}{\mathbf{\Sigma}^T \cdot \mathbf{\Sigma}^T}\mathbf{\Sigma}^T \Leftrightarrow \\ \Leftrightarrow \mathbf{F}_1 \cdot \mathbf{\Sigma}^T = \mathbf{F}_2 \cdot \mathbf{\Sigma}^T . \end{aligned}$$

□

Lemma 5.3. *Let Σ_{12} be a differentiable curve in $\mathbb{R}^2$. Let $\mathbf{x}$ be a point of Σ_{12}. Let $\mathbf{\Sigma}^N(\mathbf{x})$ be a non-zero normal vector to Σ_{12} at $\mathbf{x}$. Let the vectors*

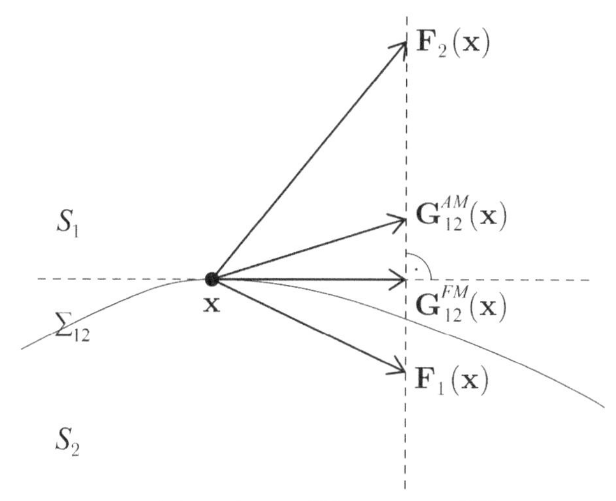

Figure 5.6 The vectors $\mathbf{F}_1(\mathbf{x})$, $\mathbf{F}_2(\mathbf{x})$, $\mathbf{G}_{12}^{AM}(\mathbf{x})$, and $\mathbf{G}_{12}^{FM}(\mathbf{x})$ at the point $\mathbf{x}$ on the differentiable curve Σ_{12}.

$$\begin{array}{l} \mathbf{F}_1(\mathbf{x})\,, \\ \mathbf{F}_2(\mathbf{x})\,, \\ \mathbf{G}_{12}^{AM}(\mathbf{x}) = \frac{1}{2}\left(\mathbf{F}_1(\mathbf{x}) + \mathbf{F}_2(\mathbf{x})\right)\,, \\ \mathbf{G}_{12}^{FM}(\mathbf{x}) = \frac{\mathbf{\Sigma}^N(\mathbf{x})\cdot\mathbf{F}_2(\mathbf{x})}{\mathbf{\Sigma}^N(\mathbf{x})\cdot(\mathbf{F}_2(\mathbf{x})-\mathbf{F}_1(\mathbf{x}))}\mathbf{F}_1(\mathbf{x}) - \frac{\mathbf{\Sigma}^N(\mathbf{x})\cdot\mathbf{F}_1(\mathbf{x})}{\mathbf{\Sigma}^N(\mathbf{x})\cdot(\mathbf{F}_2(\mathbf{x})-\mathbf{F}_1(\mathbf{x}))}\mathbf{F}_2(\mathbf{x}) \end{array}$$

have orthogonal projections $\mathrm{proj}_{\mathbf{\Sigma}^T}\mathbf{F}_1(\mathbf{x})$, $\mathrm{proj}_{\mathbf{\Sigma}^T}\mathbf{F}_2(\mathbf{x})$, $\mathrm{proj}_{\mathbf{\Sigma}^T}\mathbf{G}_{12}^{AM}(\mathbf{x})$, $\mathrm{proj}_{\mathbf{\Sigma}^T}\mathbf{G}_{12}^{FM}(\mathbf{x})$ *into the tangent line to* Σ_{12} *at* $\mathbf{x}$.

If the first and the second orthogonal projections are identical, $\mathrm{proj}_{\mathbf{\Sigma}^T}\mathbf{F}_1(\mathbf{x}) = \mathrm{proj}_{\mathbf{\Sigma}^T}\mathbf{F}_2(\mathbf{x})$, *then all defined orthogonal projections are identical,* $\mathrm{proj}_{\mathbf{\Sigma}^T}\mathbf{F}_1(\mathbf{x}) = \mathrm{proj}_{\mathbf{\Sigma}^T}\mathbf{F}_2(\mathbf{x}) = \mathrm{proj}_{\mathbf{\Sigma}^T}\mathbf{G}_{12}^{AM}(\mathbf{x}) = \mathrm{proj}_{\mathbf{\Sigma}^T}\mathbf{G}_{12}^{FM}(\mathbf{x})$.

The vectors $\mathbf{F}_1$, $\mathbf{F}_2$, $\mathbf{G}_{12}^{AM}$, and $\mathbf{G}_{12}^{FM}$ are illustrated in Figure 5.6.

Proof of Lemma 5.3:

Let $\mathrm{proj}_{\mathbf{\Sigma}^T}\mathbf{F}_1 = \mathrm{proj}_{\mathbf{\Sigma}^T}\mathbf{F}_2$. By Lemma 5.2, it holds that $\mathbf{F}_1 \cdot \mathbf{\Sigma}^T = \mathbf{F}_2 \cdot \mathbf{\Sigma}^T$. For the arithmetic mean, we have

$$\begin{aligned} \mathbf{G}_{12}^{AM} &= \tfrac{1}{2}\left(\mathbf{F}_1 + \mathbf{F}_2\right)\,, \\ \mathbf{G}_{12}^{AM} \cdot \mathbf{\Sigma}^T &= \tfrac{1}{2}\mathbf{F}_1 \cdot \mathbf{\Sigma}^T + \tfrac{1}{2}\mathbf{F}_2 \cdot \mathbf{\Sigma}^T\,, \\ \mathbf{G}_{12}^{AM} \cdot \mathbf{\Sigma}^T &= \tfrac{1}{2}\mathbf{F}_1 \cdot \mathbf{\Sigma}^T + \tfrac{1}{2}\mathbf{F}_1 \cdot \mathbf{\Sigma}^T\,, \\ \mathbf{G}_{12}^{AM} \cdot \mathbf{\Sigma}^T &= \mathbf{F}_1 \cdot \mathbf{\Sigma}^T\,, \\ \mathrm{proj}_{\mathbf{\Sigma}^T}\mathbf{G}_{12}^{AM} &= \mathrm{proj}_{\mathbf{\Sigma}^T}\mathbf{F}_1\,. \end{aligned}$$

For the Filippov method, if defined, we have

$$\begin{aligned}
\mathbf{G}_{12}^{FM} &= \frac{\mathbf{\Sigma}^N\cdot\mathbf{F}_2}{\mathbf{\Sigma}^N\cdot(\mathbf{F}_2-\mathbf{F}_1)}\mathbf{F}_1 - \frac{\mathbf{\Sigma}^N\cdot\mathbf{F}_1}{\mathbf{\Sigma}^N\cdot(\mathbf{F}_2-\mathbf{F}_1)}\mathbf{F}_2\,,\\
\mathbf{G}_{12}^{FM}\cdot\mathbf{\Sigma}^T &= \frac{\mathbf{\Sigma}^N\cdot\mathbf{F}_2}{\mathbf{\Sigma}^N\cdot(\mathbf{F}_2-\mathbf{F}_1)}\mathbf{F}_1\cdot\mathbf{\Sigma}^T - \frac{\mathbf{\Sigma}^N\cdot\mathbf{F}_1}{\mathbf{\Sigma}^N\cdot(\mathbf{F}_2-\mathbf{F}_1)}\mathbf{F}_2\cdot\mathbf{\Sigma}^T\,,\\
\mathbf{G}_{12}^{FM}\cdot\mathbf{\Sigma}^T &= \frac{\mathbf{\Sigma}^N\cdot\mathbf{F}_2}{\mathbf{\Sigma}^N\cdot(\mathbf{F}_2-\mathbf{F}_1)}\mathbf{F}_1\cdot\mathbf{\Sigma}^T - \frac{\mathbf{\Sigma}^N\cdot\mathbf{F}_1}{\mathbf{\Sigma}^N\cdot(\mathbf{F}_2-\mathbf{F}_1)}\mathbf{F}_1\cdot\mathbf{\Sigma}^T\,,\\
\mathbf{G}_{12}^{FM}\cdot\mathbf{\Sigma}^T &= \left(\frac{\mathbf{\Sigma}^N\cdot\mathbf{F}_2}{\mathbf{\Sigma}^N\cdot(\mathbf{F}_2-\mathbf{F}_1)} - \frac{\mathbf{\Sigma}^N\cdot\mathbf{F}_1}{\mathbf{\Sigma}^N\cdot(\mathbf{F}_2-\mathbf{F}_1)}\right)\mathbf{F}_1\cdot\mathbf{\Sigma}^T\,,\\
\mathbf{G}_{12}^{FM}\cdot\mathbf{\Sigma}^T &= \mathbf{F}_1\cdot\mathbf{\Sigma}^T\,,\\
\mathrm{proj}_{\mathbf{\Sigma}^T}\mathbf{G}_{12}^{FM} &= \mathrm{proj}_{\mathbf{\Sigma}^T}\mathbf{F}_1\,.
\end{aligned}$$

□

5.7 Conditions of Conservativeness

Proof of Theorem 5.1:

i) We assume that all conditions a), b), c) are satisfied. We continuously extend the vector fields $\mathbf{F}_1$ and $\mathbf{F}_2$ to $\overline{S_1}$ and $\overline{S_2}$. We denote by U_1 the potential of $\mathbf{F}_1$ on S_1 and by U_2 the potential of $\mathbf{F}_2$ on S_2. We continuously extend the potentials U_1 and U_2 to $\overline{S_1}$ and $\overline{S_2}$, which is possible because Σ_{12} is the simple boundary by Definition 5.1.

Let us consider the graph $G = (V, E)$ in Figure 5.7, where all $\triangle S_1$ and all $\triangle S_2$ are represented by the vertices V, and all $\triangle\Sigma_{12}$ are represented by the edges E. The graph G is connected because the state space S is connected. The graph G has no cycles. The state space S is simply connected and planar; so the cycle in the graph G implies the singular point on the boundary Σ_{12}, which implies the non-simple boundary by Definition 5.1. The graph G is finite or infinite, and it is the tree. From a chosen root-vertex, another vertex can be reached in exactly one possible way.

We choose any $\triangle S_1$ and any adjacent $\triangle S_2$. We take $\triangle\Sigma_{12}$ between these $\triangle S_1$ and $\triangle S_2$. We denote by $\triangle U_1$ and by $\triangle U_2$ the scalar potentials on these $\triangle S_1$ and $\triangle S_2$, respectively, on their closures $\triangle\overline{S_1}$ and $\triangle\overline{S_2}$. Let $[x_0, y_0]$ be any point in $\triangle\Sigma_{12}$. We set the constants of integration in $\triangle U_1$ and $\triangle U_2$ so that $\triangle U_1(x_0, y_0) = \triangle U_2(x_0, y_0)$. We integrate $\triangle U_1$ and $\triangle U_2$ along the curve $\triangle\Sigma_{12}$ from the point $[x_0, y_0] \in \triangle\Sigma_{12}$ to any point $[x, y] \in \triangle\Sigma_{12}$,

$$\begin{aligned}
&\triangle U_1(x,y) = \triangle U_1(x_0,y_0) + \textstyle\int \mathbf{F}_1\cdot\mathrm{d}\mathbf{r}\,,\\
&\triangle U_2(x,y) = \triangle U_2(x_0,y_0) + \textstyle\int \mathbf{F}_2\cdot\mathrm{d}\mathbf{r}\,,\quad \mathrm{d}\mathbf{r} = [\mathrm{d}x, \mathrm{d}y]\,.
\end{aligned}$$

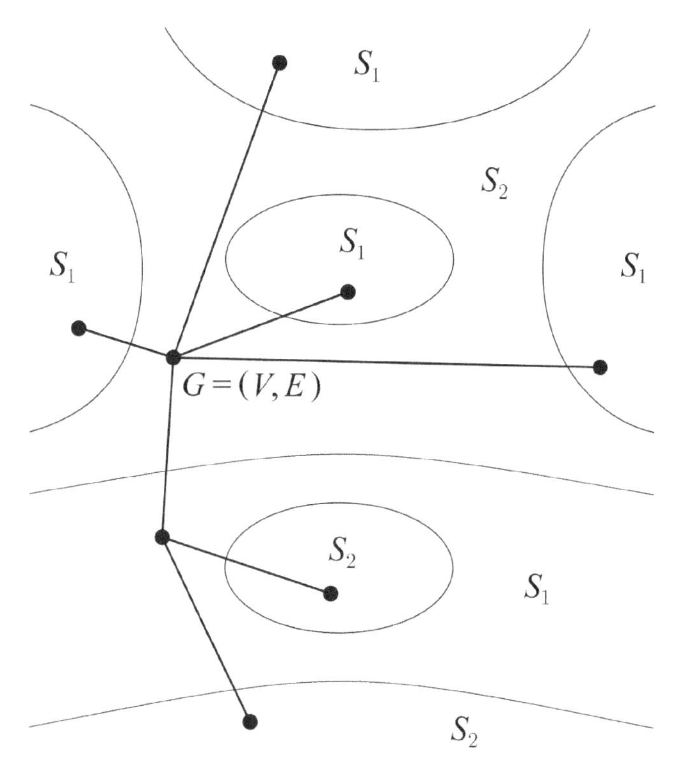

Figure 5.7 The graph $G = (V, E)$ representing the division of the simple planar Filippov state space S.

At all points of $\triangle\Sigma_{12}$, the orthogonal projections $\text{proj}_{\mathbf{\Sigma}^T}\mathbf{F}_1$ and $\text{proj}_{\mathbf{\Sigma}^T}\mathbf{F}_2$ are identical. By Lemma 5.2, we have $\mathbf{F}_1 \cdot \mathbf{\Sigma}^T = \mathbf{F}_2 \cdot \mathbf{\Sigma}^T$, where $\mathbf{\Sigma}^T$ is a non-zero tangent vector to $\triangle\Sigma_{12}$. Because $\mathbf{\Sigma}^T \| \mathrm{d}\mathbf{r}$, we have also $\mathbf{F}_1 \cdot \mathrm{d}\mathbf{r} = \mathbf{F}_2 \cdot \mathrm{d}\mathbf{r}$. These two integrals are equal; thus, $\triangle U_1(x, y) = \triangle U_2(x, y)$ for all $[x, y] \in \triangle\Sigma_{12}$.

In the same way, in accordance with the graph G in Figure 5.7, we add other connected components $\triangle S_1$ and $\triangle S_2$ till we fill the whole state space S. $\triangle U_1$ and $\triangle U_2$ on adjacent $\triangle S_1$ and $\triangle S_2$ are always equal on $\triangle\Sigma_{12}$ between them. Therefore, the scalar potential U can be continuously defined on the whole state space S,

$$U = \begin{cases} U_1 & \text{on } S_1\,, \\ U_1 = U_2 & \text{on } \Sigma_{12}\,, \\ U_2 & \text{on } S_2\,. \end{cases}$$

In the case i), the Filippov vector field $\mathbf{F}$ is conservative on the state space S.

ii) Conversely: We assume that at least one of conditions a), b), c) is not satisfied. Then three situations may arise:

ii)1) $\mathbf{F}_1$ is not conservative on some $\triangle S_1$.

ii)2) $\mathbf{F}_2$ is not conservative on some $\triangle S_2$.

ii)3) We take such $\triangle\Sigma_{12}$ and such $[x_0, y_0] \in \triangle\Sigma_{12}$ that $\mathrm{proj}_{\mathbf{\Sigma}^T}\mathbf{F}_1(x_0, y_0) \neq \mathrm{proj}_{\mathbf{\Sigma}^T}\mathbf{F}_2(x_0, y_0)$. Then there exists a neighborhood $\mathcal{O}_\epsilon(x_0, y_0) \cap \triangle\Sigma_{12}$, where $\mathrm{proj}_{\mathbf{\Sigma}^T}\mathbf{F}_1 \neq \mathrm{proj}_{\mathbf{\Sigma}^T}\mathbf{F}_2$ and $\mathbf{F}_1 \cdot \mathbf{\Sigma}^T \neq \mathbf{F}_2 \cdot \mathbf{\Sigma}^T$. We set the constants of integration in $\triangle U_1$ and $\triangle U_2$ on adjacent $\triangle\overline{S_1}$ and $\triangle\overline{S_2}$ so that $\triangle U_1(x_0, y_0) = \triangle U_2(x_0, y_0)$. We integrate $\triangle U_1$ and $\triangle U_2$ along the curve $\triangle\Sigma_{12}$ from the point $[x_0, y_0] \in \triangle\Sigma_{12}$ to any point $[x, y] \in \mathcal{O}_\epsilon(x_0, y_0) \cap \triangle\Sigma_{12} \setminus \{[x_0, y_0]\}$,

$$\begin{aligned} \triangle U_1(x,y) &= \triangle U_1(x_0,y_0) + \textstyle\int \mathbf{F}_1 \cdot \mathrm{d}\mathbf{r}\,, \\ \triangle U_2(x,y) &= \triangle U_2(x_0,y_0) + \textstyle\int \mathbf{F}_2 \cdot \mathrm{d}\mathbf{r}\,, \quad \mathrm{d}\mathbf{r} = [\mathrm{d}x, \mathrm{d}y]\,. \end{aligned}$$

On the neighborhood $\mathcal{O}_\epsilon(x_0, y_0) \cap \triangle\Sigma_{12}$, either $\mathbf{F}_1 \cdot \mathrm{d}\mathbf{r} < \mathbf{F}_2 \cdot \mathrm{d}\mathbf{r}$ or $\mathbf{F}_1 \cdot \mathrm{d}\mathbf{r} > \mathbf{F}_2 \cdot \mathrm{d}\mathbf{r}$ holds. Thus, $\triangle U_1(x, y) \neq \triangle U_2(x, y)$ on $\mathcal{O}_\epsilon(x_0, y_0) \cap \triangle\Sigma_{12} \setminus \{[x_0, y_0]\}$. Therefore, the scalar potential U can not be continuously defined on the whole state space S.

In the cases ii)1), ii)2), and ii)3), the Filippov vector field $\mathbf{F}$ is not conservative on the state space S.

□

5.8 Path Independence

Lemma 5.4. *Let M be the connected subset of $\mathbb{R}^2$. Let $\mathbf{V}$ be the vector field, $\mathbf{V} : M \to \mathbb{R}^2$. Let Σ be a differentiable curve, $\Sigma \subset M$. Let the vector field $\mathbf{W} : M \to \mathbb{R}^2$ be defined as*

$$\mathbf{W} = \begin{cases} \mathbf{V} & \text{on} \quad M \setminus \Sigma\,, \\ \mathbf{V}^* & \text{on} \quad \Sigma\,, \end{cases}$$

where $\mathbf{V}^*$ is a vector, which has the same orthogonal projection into the tangent line to Σ as the vector $\mathbf{V}$ has.

Then the curve integral (5.4) *from any point A to any point B along any piecewise differentiable curve $\mathcal{K}$ has the same value in both vector fields* $\mathbf{V}$, $\mathbf{W}$.

Proof of Lemma 5.4:

The curve integrals are

$$\int_{\mathcal{K}} \mathbf{V} \cdot \mathrm{d}\mathbf{r} \quad \text{and} \quad \int_{\mathcal{K}} \mathbf{W} \cdot \mathrm{d}\mathbf{r}\,.$$

The nonempty intersection of the differentiable curve Σ and the piecewise differentiable curve $\mathcal{K}$ has connected components, which consist either of one point (single point) or of uncountably many points (segment of curve).

i) In $M \setminus \Sigma$: Both vector fields $\mathbf{V}$, $\mathbf{W}$ are identical; so $\mathbf{V} \cdot \mathrm{d}\mathbf{r} = \mathbf{W} \cdot \mathrm{d}\mathbf{r}$. Thus, both curve integrals are identical.

ii) At the single point $\mathbf{x} \in \Sigma \cap \mathcal{K}$: The different values of the vector fields $\mathbf{V}$, $\mathbf{W}$ do not affect the values of the curve integrals.

iii) On the segment of the curve $\mathcal{X} \subset \Sigma \cap \mathcal{K}$: Let $\mathrm{proj}_{\mathbf{\Sigma}^T}\mathbf{W}$, $\mathrm{proj}_{\mathbf{\Sigma}^T}\mathbf{V}^*$, $\mathrm{proj}_{\mathbf{\Sigma}^T}\mathbf{V}$ be the orthogonal projections of the vectors $\mathbf{W}$, $\mathbf{V}^*$, $\mathbf{V}$ into the tangent line to Σ. On Σ, it holds that $\mathbf{W} = \mathbf{V}^*$ and that $\mathrm{proj}_{\mathbf{\Sigma}^T}\mathbf{V}^* = \mathrm{proj}_{\mathbf{\Sigma}^T}\mathbf{V}$. As a consequence, we have $\mathrm{proj}_{\mathbf{\Sigma}^T}\mathbf{W} = \mathrm{proj}_{\mathbf{\Sigma}^T}\mathbf{V}^* = \mathrm{proj}_{\mathbf{\Sigma}^T}\mathbf{V}$. By Lemma 5.2, it holds that $\mathbf{V} \cdot \mathbf{\Sigma}^T = \mathbf{W} \cdot \mathbf{\Sigma}^T$, where $\mathbf{\Sigma}^T$ is a non-zero tangent vector to Σ. Because $\mathbf{\Sigma}^T \| \mathrm{d}\mathbf{r}$, we have $\mathbf{V} \cdot \mathrm{d}\mathbf{r} = \mathbf{W} \cdot \mathrm{d}\mathbf{r}$. Thus, both curve integrals are identical.

Cases i), ii), and iii) show that the curve integral has the same value in both vector fields $\mathbf{V}$, $\mathbf{W}$.

□

Lemma 5.5. *Let C and D be the connected subsets of $\mathbb{R}^2$. Let the set E be $E = C \cup D$. Let the vector fields $\mathbf{C} : C \to \mathbb{R}^2$ and $\mathbf{D} : D \to \mathbb{R}^2$ be given and let their curve integrals be path independent. If the intersection $C \cap D$ is the connected set and $\mathbf{C} = \mathbf{D}$ on $C \cap D$, then the curve integral of the vector field $\mathbf{E} : E \to \mathbb{R}^2$ is path independent,*

$$\mathbf{E} = \begin{cases} \mathbf{C} & \text{on} \quad C \setminus D\,, \\ \mathbf{C} = \mathbf{D} & \text{on} \quad C \cap D\,, \\ \mathbf{D} & \text{on} \quad D \setminus C\,. \end{cases}$$

Proof of Lemma 5.5:

We choose any closed curve $\mathcal{L} \subset E$. There are four possibilities:

i) The curve $\mathcal{L}$ lies completely in the set C; thus, the curve integral along $\mathcal{L}$ is equal to zero.

ii) The curve $\mathcal{L}$ lies completely in the set D; thus, the curve integral along $\mathcal{L}$ is equal to zero.

iii) Let the oriented curve $\mathcal{L}$ start at the point $\mathbf{x}_1 \in C \setminus D$. Then it continues with its oriented part $\mathcal{L}_1$ to the point $\mathbf{x}_2 \in C \cap D$, and after that, it continues with its oriented part $\mathcal{L}_2$ to the point $\mathbf{x}_3 \in D \setminus C$. Then, it continues with its oriented part $\mathcal{L}_3$ to the point $\mathbf{x}_4 \in C \cap D$, and after that, it continues with its oriented part $\mathcal{L}_4$ to the starting point $\mathbf{x}_1 \in C \setminus D$. Let at the point $\mathbf{x}_2$ start the oriented curve $\mathcal{P}_1$, which lies completely in $C \cap D$, and which ends at the point $\mathbf{x}_4$; see Figure 5.8. The closed curve consisting of the curves $\mathcal{L}_1$, $\mathcal{P}_1$, $\mathcal{L}_4$ lies completely in C; so we have

$$\int_{\mathcal{L}_1} \mathbf{E} \cdot \mathbf{dr} + \int_{\mathcal{P}_1} \mathbf{E} \cdot \mathbf{dr} + \int_{\mathcal{L}_4} \mathbf{E} \cdot \mathbf{dr} = 0\,.$$

The closed curve consisting of the curves $\mathcal{L}_2$, $\mathcal{L}_3$, $\mathcal{P}_1$ lies completely in D; so we have

$$\int_{\mathcal{L}_2} \mathbf{E} \cdot \mathbf{dr} + \int_{\mathcal{L}_3} \mathbf{E} \cdot \mathbf{dr} - \int_{\mathcal{P}_1} \mathbf{E} \cdot \mathbf{dr} = 0\,.$$

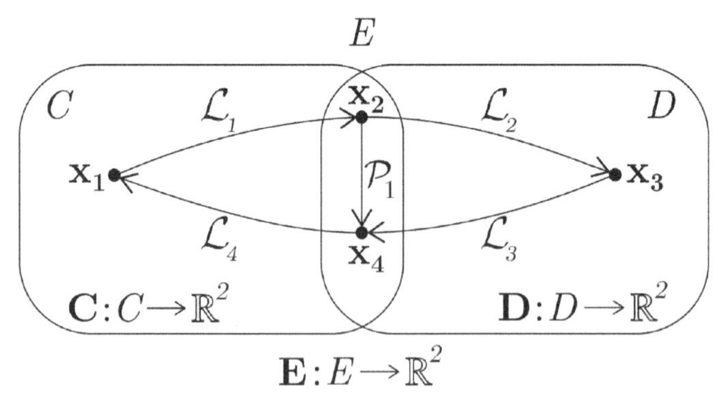

Figure 5.8 Layout of the curves $\mathcal{L}_i$ and $\mathcal{P}_i$ in the set E.

By adding two equations above, we obtain

$$\int_{\mathcal{L}_1} \mathbf{E} \cdot \mathrm{d}\mathbf{r} + \int_{\mathcal{L}_2} \mathbf{E} \cdot \mathrm{d}\mathbf{r} + \int_{\mathcal{L}_3} \mathbf{E} \cdot \mathrm{d}\mathbf{r} + \int_{\mathcal{L}_4} \mathbf{E} \cdot \mathrm{d}\mathbf{r} = 0\,.$$

Thus, the curve integral along $\mathcal{L}$ is equal to zero.

iv) The closed curve $\mathcal{L}$ is more complex. By appropriately introducing the oriented curves $\mathcal{L}_i$ and $\mathcal{P}_i$, this case can be converted to the previous cases.

In the cases i), ii), iii), and iv), the curve integral of the vector field $\mathbf{E}$ is path independent.

□

Proof of Theorem 5.2:

i) We assume that the vector field $\mathbf{F} : S \to \mathbb{R}^2$ is conservative. Then $\mathbf{F}_1 : S_1 \to \mathbb{R}^2$ is conservative with $U_1 : S_1 \to \mathbb{R}$. $\mathbf{F}_1$ can be continuously extended to $\overline{S_1}$. U_1 can be continuously extended to $\overline{S_1}$ because Σ_{12} is the simple boundary by Definition 5.1. $\mathbf{F}_1 : \overline{S_1} \to \mathbb{R}^2$ is conservative with $U_1 : \overline{S_1} \to \mathbb{R}$ by Definition 5.7. The curve integral of $\mathbf{F}_1 : \overline{S_1} \to \mathbb{R}^2$ is path independent by Lemma 5.1.

Similarly, the curve integral of $\mathbf{F}_2 : \overline{S_2} \to \mathbb{R}^2$ is path independent.

We choose any $\triangle\Sigma_{12}$. We take an adjacent $\triangle S_1$. We define the vector field $\widetilde{\mathbf{F}}_1 : \triangle\overline{S_1} \to \mathbb{R}^2$,

$$\widetilde{\mathbf{F}}_1 = \begin{cases} \mathbf{F}_1 & \text{on} \quad \triangle S_1\,, \\ \mathrm{proj}_{\mathbf{\Sigma}T}\mathbf{F}_1 & \text{on} \quad \triangle\Sigma_{12}\,, \end{cases}$$

where $\mathrm{proj}_{\mathbf{\Sigma}T}\mathbf{F}_1$ is the orthogonal projection of $\mathbf{F}_1$ into the tangent line to Σ_{12}. By Lemma 5.4, the curve integral of the vector field $\widetilde{\mathbf{F}}_1$ is path independent.
We take a $\triangle S_2$ adjacent to the chosen $\triangle\Sigma_{12}$. We define the vector field $\widetilde{\mathbf{F}}_2 : \triangle\overline{S_2} \to \mathbb{R}^2$,

$$\widetilde{\mathbf{F}}_2 = \begin{cases} \mathbf{F}_2 & \text{on} \quad \triangle S_2\,, \\ \mathrm{proj}_{\mathbf{\Sigma}T}\mathbf{F}_2 & \text{on} \quad \triangle\Sigma_{12}\,, \end{cases}$$

where $\mathrm{proj}_{\mathbf{\Sigma}T}\mathbf{F}_2$ is the orthogonal projection of $\mathbf{F}_2$ into the tangent line to Σ_{12}. By Lemma 5.4, the curve integral of the vector field $\widetilde{\mathbf{F}}_2$ is path independent.

We define the vector field $\widetilde{\mathbf{F}} : \triangle S_1 \cup \triangle\Sigma_{12} \cup \triangle S_2 \to \mathbb{R}^2$,

$$\widetilde{\mathbf{F}} = \begin{cases} \mathbf{F}_1 & \text{on} \quad \triangle S_1\,, \\ \mathrm{proj}_{\mathbf{\Sigma}^T}\mathbf{F}_1 = \mathrm{proj}_{\mathbf{\Sigma}^T}\mathbf{F}_2 & \text{on} \quad \triangle\Sigma_{12}\,, \\ \mathbf{F}_2 & \text{on} \quad \triangle S_2\,, \end{cases}$$

where $\triangle S_1$, $\triangle\Sigma_{12}$, $\triangle S_2$ are adjacent to each other. By Lemma 5.5, the curve integral of the vector field $\widetilde{\mathbf{F}}$ is path independent.

In the same way, in accordance with the graph G in Figure 5.7, we add other connected components $\triangle S_1$ and $\triangle S_2$ till we fill the whole state space S. We obtain the vector field $\widetilde{\mathbf{F}} : S \to \mathbb{R}^2$,

$$\widetilde{\mathbf{F}} = \begin{cases} \mathbf{F}_1 & \text{on} \quad S_1\,, \\ \mathrm{proj}_{\mathbf{\Sigma}^T}\mathbf{F}_1 = \mathrm{proj}_{\mathbf{\Sigma}^T}\mathbf{F}_2 & \text{on} \quad \Sigma_{12}\,, \\ \mathbf{F}_2 & \text{on} \quad S_2\,. \end{cases}$$

The curve integral of the vector field $\widetilde{\mathbf{F}}$ is path independent. Using Lemma 5.3, we turn it into the vector field $\widehat{\mathbf{F}} : S \to \mathbb{R}^2$,

$$\widehat{\mathbf{F}} = \begin{cases} \mathbf{F}_1 & \text{on} \quad S_1\,, \\ \mathbf{F}_2 & \text{on} \quad S_2\,, \\ \mathrm{proj}_{\mathbf{\Sigma}^T}\mathbf{G}_{12}^{AM} & \text{on} \quad \Sigma_{12}^{C} \cup \Sigma_{12}^{PD}\,, \\ \mathrm{proj}_{\mathbf{\Sigma}^T}\mathbf{G}_{12}^{FM} & \text{on} \quad \Sigma_{12}^{S} \setminus \Sigma_{12}^{PD}. \end{cases}$$

The curve integral of the vector field $\widehat{\mathbf{F}}$ is path independent. Using Lemma 5.4, we turn it into the vector field $\mathbf{F} : S \to \mathbb{R}^2$,

$$\mathbf{F} = \begin{cases} \mathbf{F}_1 & \text{on} \quad S_1\,, \\ \mathbf{F}_2 & \text{on} \quad S_2\,, \\ \mathbf{G}_{12}^{AM} & \text{on} \quad \Sigma_{12}^{C} \cup \Sigma_{12}^{PD}\,, \\ \mathbf{G}_{12}^{FM} & \text{on} \quad \Sigma_{12}^{S} \setminus \Sigma_{12}^{PD}\,, \end{cases}$$

which is the simple planar Filippov vector field

$$\mathbf{F} = \begin{cases} \mathbf{F}_1 & \text{on} \quad S_1\,, \\ \mathbf{F}_2 & \text{on} \quad S_2\,, \\ \mathbf{G}_{12} & \text{on} \quad \Sigma_{12}\,. \end{cases}$$

In the case i), the curve integral of the simple planar Filippov vector field $\mathbf{F}$ is path independent.

ii) Conversely: We assume that the vector field $\mathbf{F} : S \to \mathbb{R}^2$ is not conservative. Then three situations may arise:

ii)1) $\mathbf{F}_1$ is not conservative on some $\triangle S_1$.

ii)2) $\mathbf{F}_2$ is not conservative on some $\triangle S_2$.

ii)3) There exists a point $[x_0, y_0]$ in some $\triangle\Sigma_{12}$, and there exist constants of integration such that $U_1(x_0, y_0) = U_2(x_0, y_0)$, but $U_1(x, y) \neq U_2(x, y)$ in reduced neighborhood $\mathcal{O}^R_\epsilon(x_0, y_0)$ laying on Σ_{12}. Let the point $A = [x_0, y_0]$ be in $\triangle\Sigma_{12}$ and let the point $B = [x, y]$ be in $\mathcal{O}^R_\epsilon(x_0, y_0)$ on Σ_{12}. We choose two oriented curves $\widetilde{\mathcal{K}}$ and $\widehat{\mathcal{K}}$ in adjacent $\triangle\overline{S_1}$ and $\triangle\overline{S_2}$, $\widetilde{\mathcal{K}} \subset \triangle\overline{S_1}$, $\widehat{\mathcal{K}} \subset \triangle\overline{S_2}$. They start at the same point A, and they end at the same point B, but they follow different paths. The curve integrals along the curves $\widetilde{\mathcal{K}}$ and $\widehat{\mathcal{K}}$ give

$$\begin{aligned} U_1(B) &= U_1(A) + \textstyle\int_{\widetilde{\mathcal{K}}} \mathbf{F} \cdot \mathrm{d}\mathbf{r}\,, \\ U_2(B) &= U_2(A) + \textstyle\int_{\widehat{\mathcal{K}}} \mathbf{F} \cdot \mathrm{d}\mathbf{r}\,, \quad \mathrm{d}\mathbf{r} = [\mathrm{d}x, \mathrm{d}y]\,. \end{aligned}$$

Because $U_1(A) = U_2(A)$ and $U_1(B) \neq U_2(B)$, the curve integrals along $\widetilde{\mathcal{K}}$ and $\widehat{\mathcal{K}}$ have different values.

In the cases ii)1), ii)2), and ii)3), the curve integral of the simple planar Filippov vector field $\mathbf{F}$ is path dependent.

□

5.9 Trajectories and Fall Lines

In general, Filippov systems possess no flow $\varphi : X \times \mathbb{R} \to X$. They are characterized by the existence, but not by the uniqueness, of a solution through a given point.

The trajectory of a simple planar Filippov system is a piecewise differentiable curve in a simple planar Filippov state space S. The trajectory has one connected component. The trajectory has regular points almost everywhere except for some points: common equilibrium CE, boundary equilibrium BE, crossing point CP, onset point ONP, offset point OFP. These points together with tangent points TP cut trajectories into segments: subtrajectory in S_1, subtrajectory in S_2, sliding subtrajectory. Uncountably many sliding subtrajectories overlap on a sliding set Σ^S_{12}. They can be understood as a bundle of sliding subtrajectories or as a single sliding subtrajectory.

The sliding set Σ_{12}^S consists of accumulated onset points ONP, and offset points OFP. Each ONP, and each OFP connects a sliding subtrajectory to a subtrajectory in S_1 or in S_2 by a spike joint. In the case of stable sliding, the subtrajectory in S_1 or in S_2 reaches Σ_{12}^S at ONP, and after that, it follows Σ_{12}^S as the sliding subtrajectory. In the case of unstable sliding, the sliding subtrajectory follows Σ_{12}^S. Then it leaves Σ_{12}^S at OFP, and after that, it continues as the subtrajectory in S_1 or in S_2. The sliding set Σ_{12}^S also contains boundary equilibria BE and tangent points TP. Each BE connects itself, understood as a sliding subtrajectory, to a subtrajectory in S_1 or in S_2. Each TP connects a sliding subtrajectory to a subtrajectory in S_1 or in S_2 by a smooth joint.

The crossing set Σ_{12}^C consists of accumulated crossing points CP. Each CP connects a subtrajectory in S_1 to a subtrajectory in S_2 by a spike joint. In the case of crossing, the subtrajectory in S_1 reaches Σ_{12}^C at CP, and after that, it continues as the subtrajectory in S_2. Or conversely, it comes from S_2 and goes to S_1.

A phase portrait of a simple planar Filippov system is illustrated in Figure 5.9.

Proof of Theorem 5.3:

i) In the union S_1: The subtrajectories in S_1 are led by the continuous vector field $\mathbf{F}_1$. The fall sublines in S_1 are guided by the continuous vector field ∇U_1. Identical directional vector fields, $\mathbf{F}_1 = \nabla U_1$, give identical systems of curves: subtrajectories and fall sublines.

ii) In the union S_2: The subtrajectories in S_2 are led by the continuous vector field $\mathbf{F}_2$. The fall sublines in S_2 are guided by the continuous vector field ∇U_2. Identical directional vector fields, $\mathbf{F}_2 = \nabla U_2$, give identical systems of curves: subtrajectories and fall sublines.

iii) On the crossing set Σ_{12}^C: There are no subtrajectories on Σ_{12}^C.

iv) On the sliding set Σ_{12}^S: The sliding subtrajectory is directed by the sliding set Σ_{12}^S, respectively, by the vector field $\mathbf{G}_{12}$.

At one point $[\mathbf{x}, U(\mathbf{x})]$, $\mathbf{x} \in \Sigma_{12}^S$, the topographic surface U has two tangent planes ρ_1 and ρ_2; see Figure 5.10. Their normal vectors are

$$\widehat{\mathbf{W}}_{\rho_1} = \left[\frac{\partial U_1}{\partial x}, \frac{\partial U_1}{\partial y}, -1\right] \quad \text{and} \quad \widehat{\mathbf{W}}_{\rho_2} = \left[\frac{\partial U_2}{\partial x}, \frac{\partial U_2}{\partial y}, -1\right].$$

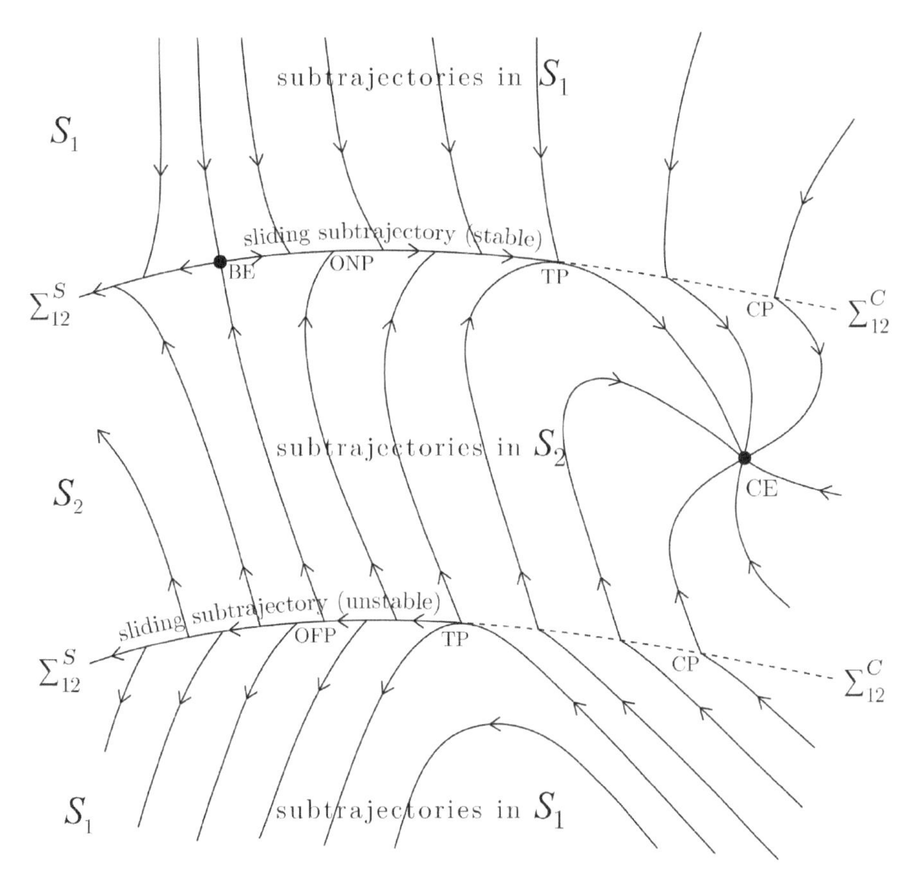

Figure 5.9 Example of a phase portrait of a simple planar Filippov system. Important points: common equilibrium CE, boundary equilibrium BE, tangent point TP, crossing point CP, onset point ONP, and offset point OFP.

At the points with two different tangent planes, the topographic surface U has the fracture curve. The tangent vector to the fracture curve is the cross product

$$\widehat{\mathbf{W}} = \widehat{\mathbf{W}}_{\rho_1} \times \widehat{\mathbf{W}}_{\rho_2} .$$

In terms of partial derivatives, the vector $\widehat{\mathbf{W}}$ is

$$\widehat{\mathbf{W}} = [\widehat{W}_x, \widehat{W}_y, \widehat{W}_z] = \\ = \left[-\frac{\partial U_1}{\partial y} + \frac{\partial U_2}{\partial y}, \frac{\partial U_1}{\partial x} - \frac{\partial U_2}{\partial x}, \frac{\partial U_1}{\partial x}\frac{\partial U_2}{\partial y} - \frac{\partial U_1}{\partial y}\frac{\partial U_2}{\partial x} \right] .$$

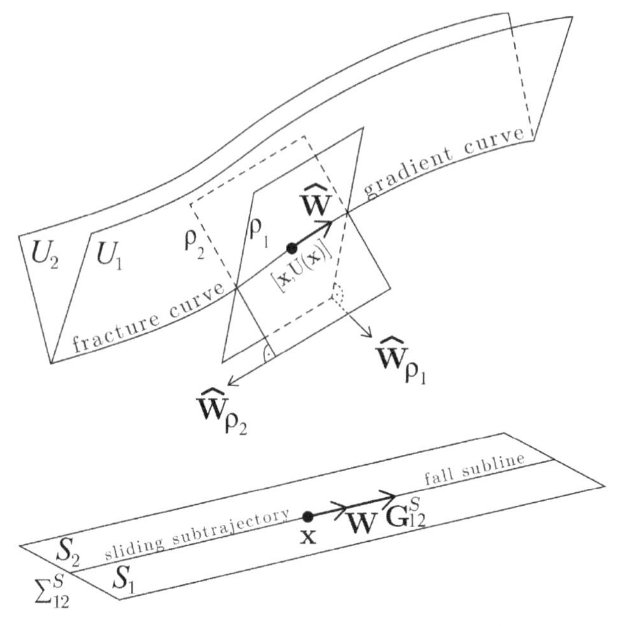

Figure 5.10 The topographic surface U is displayed above the horizontal plane S.

The orthogonal projection of the vector $\widehat{\mathbf{W}}$ into the horizontal plane S is

$$\mathbf{W} = [W_x, W_y] = \left[-\frac{\partial U_1}{\partial y} + \frac{\partial U_2}{\partial y}, \frac{\partial U_1}{\partial x} - \frac{\partial U_2}{\partial x}\right].$$

In terms of the vectors $\mathbf{F}_1 = [F_{1x}, F_{1y}]$, $\mathbf{F}_2 = [F_{2x}, F_{2y}]$, the vector $\mathbf{W}$ is

$$\mathbf{W} = [-F_{1y} + F_{2y},\ F_{1x} - F_{2x}] = [-F_{1y}, F_{1x}] + [F_{2y}, -F_{2x}] = \curvearrowleft \mathbf{F}_1 + \mathbf{F}_2 \curvearrowright,$$

where $\curvearrowleft$ is counterclockwise 90-degree turn of vector, and $\curvearrowright$ is clockwise 90-degree turn of vector.

The simple planar Filippov vector field $\mathbf{F}$ is conservative; thus, by Theorem 5.1, and by Lemma 5.2, we have

$$\begin{aligned} \mathrm{proj}_{\mathbf{\Sigma}^T}\mathbf{F}_2 &= \mathrm{proj}_{\mathbf{\Sigma}^T}\mathbf{F}_1\,, \\ \mathbf{F}_2 \cdot \mathbf{\Sigma}^T &= \mathbf{F}_1 \cdot \mathbf{\Sigma}^T\,, \\ \mathbf{F}_2 \cdot \mathbf{\Sigma}^T - \mathbf{F}_1 \cdot \mathbf{\Sigma}^T &= 0\,, \\ (\mathbf{F}_2 - \mathbf{F}_1) \cdot \mathbf{\Sigma}^T &= 0\,. \end{aligned}$$

iv)1) On the sliding set, except for the points with parallel difference, $\Sigma_{12}^{S} \setminus \Sigma_{12}^{PD}$: It holds that $\mathbf{F}_1 \neq \mathbf{F}_2$ and that $\mathbf{F}_1 \nparallel \mathbf{F}_2$. The line of greatest slope (gradient curve) is directed by the fracture curve of the topographic surface U, respectively, by the vector field $\widehat{\mathbf{W}}$. The fall subline is driven by the vector field $\mathbf{W}$. In this case, we can take $\mathbf{\Sigma}^N = \mathbf{F}_2 - \mathbf{F}_1$ as the non-zero normal vector to Σ_{12}^{S}. Then

$$\mathbf{G}_{12} = \frac{(\mathbf{F}_2 - \mathbf{F}_1) \cdot \mathbf{F}_2}{(\mathbf{F}_2 - \mathbf{F}_1) \cdot (\mathbf{F}_2 - \mathbf{F}_1)} \mathbf{F}_1 - \frac{(\mathbf{F}_2 - \mathbf{F}_1) \cdot \mathbf{F}_1}{(\mathbf{F}_2 - \mathbf{F}_1) \cdot (\mathbf{F}_2 - \mathbf{F}_1)} \mathbf{F}_2 \,.$$

Because the determinant $\begin{vmatrix} \mathbf{W} \\ \mathbf{G}_{12} \end{vmatrix}$ is a constant zero,

$$\begin{vmatrix} \mathbf{W} \\ \mathbf{G}_{12} \end{vmatrix} = 0 \quad \forall \mathbf{x} \in \Sigma_{12}^{S} \setminus \Sigma_{12}^{PD} \,,$$

the vectors $\mathbf{W}$ and $\mathbf{G}_{12}$ are always linearly dependent. Thus, $\mathbf{G}_{12}$ is a real multiple of $\mathbf{W}$,

$$\mathbf{G}_{12} = c\,\mathbf{W} \,, \quad c \in \mathbb{R} \,, \quad \mathbf{W} \neq \mathbf{0} \,.$$

At one point of Σ_{12}^{S}, c is a constant of multiplication, but along Σ_{12}^{S}, c is a variable. We find the formula for c as follows:

$$c = c_1 c_2 = \frac{\widehat{W}_z}{\sqrt{\widehat{W}_x^2 + \widehat{W}_y^2}} \frac{1}{\|\mathbf{W}\|} = \frac{\widehat{W}_z}{\sqrt{W_x^2 + W_y^2}} \frac{1}{\|\mathbf{W}\|} = \frac{\widehat{W}_z}{\|\mathbf{W}\|} \frac{1}{\|\mathbf{W}\|} =$$

$$= \frac{\widehat{W}_z}{\|\mathbf{W}\|^2} = \frac{\dfrac{\partial U_1}{\partial x} \dfrac{\partial U_2}{\partial y} - \dfrac{\partial U_1}{\partial y} \dfrac{\partial U_2}{\partial x}}{\|\mathbf{W}\|^2} = \frac{F_{1x} F_{2y} - F_{1y} F_{2x}}{\|\mathbf{W}\|^2} =$$

$$= \frac{\begin{vmatrix} \mathbf{F}_1 \\ \mathbf{F}_2 \end{vmatrix}}{\| \frown \mathbf{F}_1 + \mathbf{F}_2 \curvearrowright \|^2} \,.$$

iv)2) On the sliding set, in the case of double tangency, Σ_{12}^{DT}: It holds that $\mathbf{F}_1 = \mathbf{F}_2$ and that $\mathbf{F}_1 \parallel \mathbf{F}_2$. The planes ρ_1, ρ_2 merge into one plane ρ, and the fracture curve vanishes as well as the vectors, $\widehat{\mathbf{W}} = \mathbf{0}$, $\mathbf{W} = \mathbf{0}$. The representant of the greatest slope of the plane ρ, respectively, of the topographic surface U is the gradient, $\nabla U = \mathbf{F}_1 = \mathbf{F}_2$. In this case, $\mathbf{G}_{12}$ is defined by the arithmetic mean of $\mathbf{F}_1$ and $\mathbf{F}_2$; so $\mathbf{G}_{12} = \mathbf{F}_1 = \mathbf{F}_2$. Thus, $\mathbf{G}_{12}$ is equal to ∇U,

$$\mathbf{G}_{12} = \nabla U \,, \quad \mathbf{W} = \mathbf{0} \,.$$

iv) The vector $\mathbb{W}$ is the sum of the vectors from cases iv)1), and iv)2),

$$\mathbb{W} = \begin{cases} c\,\mathbf{W}\,, & \mathbf{W} \neq \mathbf{0}\,, \\ \nabla U\,, & \mathbf{W} = \mathbf{0}\,. \end{cases}$$

In terms of the vectors $\mathbf{F}_1$, $\mathbf{F}_2$, the vector $\mathbb{W}$ is

$$\mathbb{W} = \begin{cases} \dfrac{\begin{vmatrix} \mathbf{F}_1 \\ \mathbf{F}_2 \end{vmatrix}}{\| \curvearrowleft \mathbf{F}_1 + \mathbf{F}_2 \curvearrowright \|^2} (\curvearrowleft \mathbf{F}_1 + \mathbf{F}_2 \curvearrowright)\,, & \mathbf{F}_1 - \mathbf{F}_2 \neq \mathbf{0}\,, \\ \\ \mathbf{F}_1\,, & \mathbf{F}_1 - \mathbf{F}_2 = \mathbf{0}\,. \end{cases}$$

The sliding subtrajectories on Σ_{12}^S are led by the vector field $\mathbf{G}_{12}$. The fall sublines on Σ_{12}^S are guided by the vector field $\mathbb{W}$. Identical directional vector fields, $\mathbf{G}_{12} = \mathbb{W}$, give identical systems of curves: sliding subtrajectories and fall sublines.

Cases i), ii), iii), and iv) show that to a given subtrajectory always exists an identical fall subline and vice versa.

The whole trajectory can be disassembled into its subtrajectories. These subtrajectories can be converted into identical fall sublines. These fall sublines can be assembled in a whole fall line. The process can be done in exactly one possible way. To the given whole trajectory there is always an identical whole fall line and vice versa. □

5.10 Examples

Example 5.1. Two vector fields $\mathbf{F}_1$, $\mathbf{F}_2$ and one scalar function h are given:

$$\mathbf{F}_1(x, y) = \begin{bmatrix} -x \\ -y - 1 \end{bmatrix}, \quad \mathbf{F}_2(x, y) = \begin{bmatrix} -x \\ -y + 1 \end{bmatrix}, \quad h(x, y) = y\,.$$

These functions give the simple planar Filippov vector field:

$$\mathbf{F}(x, y) = \begin{cases} \begin{bmatrix} -x \\ -y - 1 \end{bmatrix} & y > 0\,,\ x \in \mathbb{R}\,, \\ \begin{bmatrix} -x \\ -y + 1 \end{bmatrix} & y < 0\,,\ x \in \mathbb{R}\,, \\ \begin{bmatrix} -x \\ -y \end{bmatrix} & y = 0\,,\ y^2 - 1 > 0\,,\ x \in \mathbb{R}\,, \\ \begin{bmatrix} -x \\ 0 \end{bmatrix} & y = 0\,,\ y^2 - 1 \leq 0\,,\ x \in \mathbb{R}\,. \end{cases}$$

In this vector field, there are no points with the double tangency. The conditions $y = 0$, $y^2 - 1 > 0$ have no solution. In this vector field, there is empty crossing set. The conditions $y = 0$, $y^2 - 1 \leq 0$ give only $y = 0$. So we can simplify the notation of this vector field to

$$\mathbf{F}(x,y) = \begin{cases} \begin{bmatrix} -x \\ -y-1 \end{bmatrix} & y > 0\,,\ x \in \mathbb{R}\,, \\ \begin{bmatrix} -x \\ -y+1 \end{bmatrix} & y < 0\,,\ x \in \mathbb{R}\,, \\ \begin{bmatrix} -x \\ 0 \end{bmatrix} & y = 0\,,\ x \in \mathbb{R}\,. \end{cases} \tag{5.5}$$

The scalar function U is given:

$$U(x,y) = \begin{cases} -\frac{1}{2}x^2 - \frac{1}{2}(y+1)^2\,, & y \geq 0\,,\ x \in \mathbb{R}\,, \\ -\frac{1}{2}x^2 - \frac{1}{2}(y-1)^2\,, & y < 0\,,\ x \in \mathbb{R}\,. \end{cases} \tag{5.6}$$

The function U fulfills the conditions in Definition 5.6. Let us check it.

a) For $y > 0$: $\quad \dfrac{\partial U}{\partial x} = -x, \quad \dfrac{\partial U}{\partial y} = -y - 1.$

b) For $y < 0$: $\quad \dfrac{\partial U}{\partial x} = -x, \quad \dfrac{\partial U}{\partial y} = -y + 1.$

c) For $y = 0$: $\quad -\dfrac{1}{2}x^2 - \dfrac{1}{2}(y+1)^2 = -\dfrac{1}{2}x^2 - \dfrac{1}{2}$

and

$$-\frac{1}{2}x^2 - \frac{1}{2}(y-1)^2 = -\frac{1}{2}x^2 - \frac{1}{2}.$$

Thus, the simple planar Filippov vector field $\mathbf{F}$ is conservative and the function U is its scalar potential; see Figure 5.11.

Example 5.2. Two vector fields $\mathbf{F}_1$, $\mathbf{F}_2$ and one scalar function h are given:

$$\mathbf{F}_1(x,y) = \begin{bmatrix} xy \\ \frac{1}{2}x^2 + 2y \end{bmatrix}, \quad \mathbf{F}_2(x,y) = \begin{bmatrix} y \\ x + 4y - 1 \end{bmatrix},$$
$$h(x,y) = -x^2 + 2x + 2y - 2\,.$$

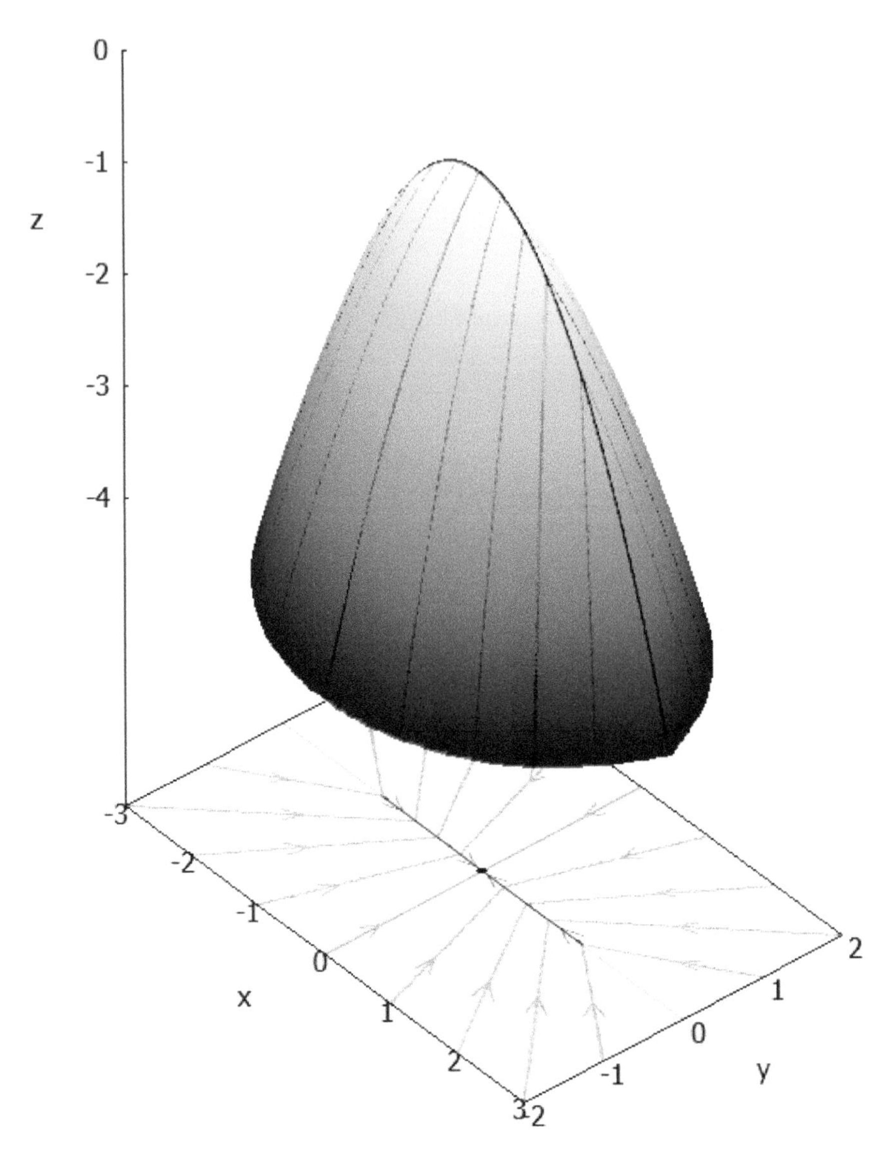

Figure 5.11 The graph of the scalar potential U in equation (5.6) is displayed above the phase portrait of the simple planar Filippov system given by the simple planar Filippov vector field $\mathbf{F}$ in equation (5.5). On the graph, the lines of greatest slope are depicted. The values of U are on the z-axis.

These functions give the simple planar Filippov vector field:

$$\mathbf{F}(x,y) = \begin{cases} \begin{bmatrix} xy \\ \frac{1}{2}x^2 + 2y \end{bmatrix} & -x^2 + 2x + 2y - 2 > 0\,, \\ \begin{bmatrix} y \\ x + 4y - 1 \end{bmatrix} & -x^2 + 2x + 2y - 2 < 0\,, \\ \begin{bmatrix} \frac{1}{2}xy + \frac{1}{2}y \\ \frac{1}{4}x^2 + \frac{1}{2}x + 3y - \frac{1}{2} \end{bmatrix} & \begin{array}{l} -x^2 + 2x + 2y - 2 = 0\,, \\ w > 0\,, \end{array} \\ \begin{bmatrix} \dfrac{-\frac{1}{2}x^2y - 4xy^2 + xy + 2y^2}{-x^2y + 2xy + \frac{1}{2}x^2 - x - 3y + 1} \\ \dfrac{-4x^2y^2 - \frac{1}{2}x^3y + \frac{3}{2}x^2y + 6xy^2 - xy - 2y^2}{-x^2y + 2xy + \frac{1}{2}x^2 - x - 3y + 1} \end{bmatrix} & \begin{array}{l} -x^2 + 2x + 2y - 2 = 0\,, \\ w \leq 0, \end{array} \end{cases} \tag{5.7}$$

where $w = \left(\frac{1}{2}x^4 - \frac{3}{2}x^3 + \frac{1}{2}x^2 + x - 2\right)\left(\frac{1}{2}x^3 - \frac{7}{2}x^2 + 5x - 4\right)$.

The vector fields $\mathbf{F}_1$ and $\mathbf{F}_2$ together with the boundary Σ_{12} fulfill the conditions in Theorem 5.1. Let us check it.

a) For $\mathbf{F}_1$: $\dfrac{\partial F_{1x}}{\partial y} = x\,, \qquad \dfrac{\partial F_{1y}}{\partial x} = x\,.$

b) For $\mathbf{F}_2$: $\dfrac{\partial F_{2x}}{\partial y} = 1\,, \qquad \dfrac{\partial F_{2y}}{\partial x} = 1\,.$

c) The boundary Σ_{12} is the parabola $P\,:\, -x^2 + 2x + 2y - 2 = 0$. We parameterize it: $\mathbf{P}(t) = [t\,,\, \frac{1}{2}t^2 - t + 1]$, $t \in \mathbb{R}$. The non-zero tangent vector $\mathbf{\Sigma}^T$ is $\mathbf{P}'(t) = [1\,,\, t-1]$, $t \in \mathbb{R}$.

The vectors $\mathbf{F}_1$ and $\mathbf{F}_2$ on the boundary $\Sigma_{12} = P$ are

$$\mathbf{F}_1(t) = \left[\frac{1}{2}t^3 - t^2 + t\,, \frac{3}{2}t^2 - 2t + 2\right],$$

$$\mathbf{F}_2(t) = \left[\frac{1}{2}t^2 - t + 1\,, 2t^2 - 3t + 3\right].$$

Their orthogonal projections into the tangent line are

$$\mathrm{proj}_{\mathbf{\Sigma}^T}\mathbf{F}_1 = \mathrm{proj}_{\mathbf{P}'}\mathbf{F}_1 = \frac{\mathbf{F}_1 \cdot \mathbf{P}'}{\mathbf{P}' \cdot \mathbf{P}'} \cdot \mathbf{P}' = \frac{2t^3 - \frac{9}{2}t^2 + 5t - 2}{t^2 - 2t + 2} \cdot [1\,,\, t-1]\,,$$

$$\mathrm{proj}_{\mathbf{\Sigma}^T}\mathbf{F}_2 = \mathrm{proj}_{\mathbf{P}'}\mathbf{F}_2 = \frac{\mathbf{F}_2 \cdot \mathbf{P}'}{\mathbf{P}' \cdot \mathbf{P}'} \cdot \mathbf{P}' = \frac{2t^3 - \frac{9}{2}t^2 + 5t - 2}{t^2 - 2t + 2} \cdot [1\,,\, t-1]\,.$$

Thus, the simple planar Filippov vector field $\mathbf{F}$ is conservative. Let us compute its scalar potential U.

For $\mathbf{F}_1$: $\qquad U_1 = \frac{1}{2}x^2y + y^2 + C_1 \, .$

For $\mathbf{F}_2$: $\qquad U_2 = xy + 2y^2 - y + C_2 \, .$

We adjust the constants of integration C_1 and C_2 so that U_1 and U_2 meet continuously on the boundary $\Sigma_{12} = P$. We obtain the scalar potential U of the simple planar Filippov vector field $\mathbf{F}$; see Figure 5.12.

$$U(x, y) = \begin{cases} \frac{1}{2}x^2y + y^2 \, , & y \geq \frac{1}{2}x^2 - x + 1 \, , \\ xy + 2y^2 - y \, , & y < \frac{1}{2}x^2 - x + 1 \, . \end{cases} \tag{5.8}$$

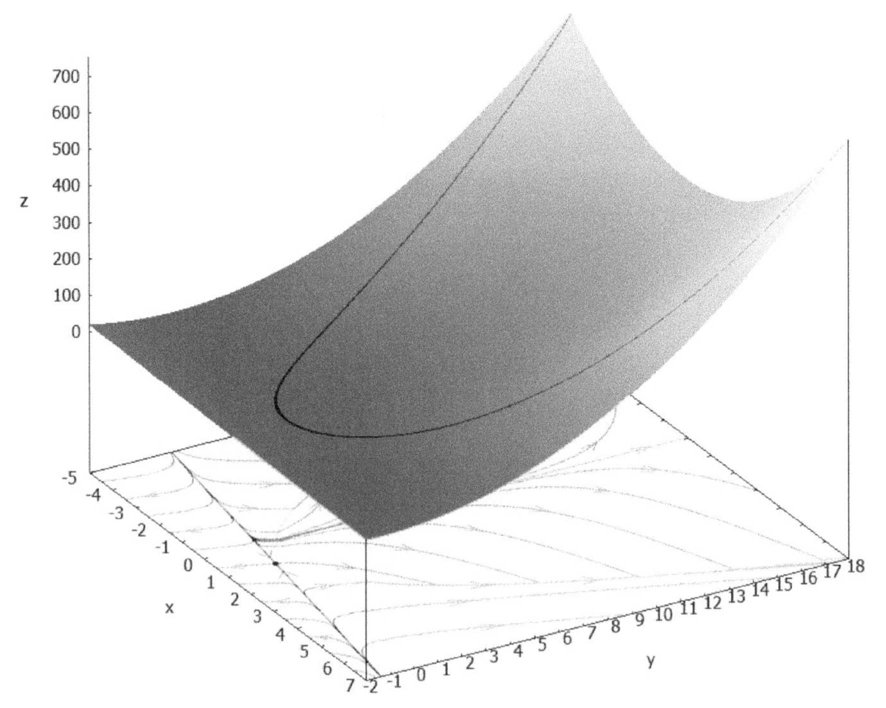

Figure 5.12 The graph of the scalar potential U in equation (5.8) is displayed above the phase portrait of the simple planar Filippov system given by the simple planar Filippov vector field $\mathbf{F}$ in equation (5.7). The values of U are on the z-axis.

5.11 Conclusion

In this chapter, we focused on piecewise smooth dynamical systems; in particular, we studied Filippov systems. We have shown that the scalar potential can be defined even if the vector field is discontinuous at some points.

Our main contributions are formulations and proofs of theorems about the existence and properties of a scalar potential in a simple planar Filippov system.

On the one hand, the conservative vector fields form only a zero-measure subset of all vector fields. On the other hand, among the important vector fields (continuous and discontinuous), there are many conservative ones.

The scalar potential facilitates the calculation of curve integrals. The scalar potential carries complete information about the global behavior of a dynamical system because the system of fall lines is also a system of trajectories.

References

[1] D. V. Anosov, S. K. Aranson, V. I. Arnold, I. U. Bronshtein, V. Z. Grines, Y. S. Ilyashenko, 'Dynamical systems I', Springer-Verlag, 1988.
[2] V. I. Babitsky, V. L. Krupenin, 'Vibration of strongly nonlinear discontinuous systems', Springer-Verlag, 2001.
[3] S. Banerjee, C. Grebogi, 'Border collision bifurcations in two-dimensional piecewise smooth maps', Physical Rewiew E, 59, 1999, pp. 4052–4061.
[4] A. D. Bazykin, 'Matematicheskaja biofizika vzaimodejstvujushchich populjacij' (in Russian), Nauka Publishing House, 1985.
[5] M. di Bernardo, C. J. Budd, A. R. Champneys, P. Kowalczyk, 'Piecewise-smooth dynamical systems, theory and applications', Springer-Verlag, 2008.
[6] B. Brogliato, 'Nonsmooth mechanics - models, dynamics and control', Springer-Verlag, 1999.
[7] C. Budd, P. Piiroinen, 'Corner bifurcations in non-smoothly forced impact oscillators', Physica D, 220, 2006, pp. 127–145.
[8] H. Dankowicz, P. Piiroinen, A. B. Nordmark, 'Low-velocity impacts of quasiperiodic oscillations', Chaos, solitons & fractals, 14 (2), 2002, pp. 241–255.

[9] F. Dercole, A. Gragnani, S. Rinaldi, 'Bifurcation analysis of piecewise smooth ecological models', Theoretical Population Biology, 72, 2007, pp. 197-213.

[10] F. Dumortier, J. Llibre, J. C. Artés, 'Qualitative theory of planar differential systems', Springer-Verlag, 2006.

[11] A. F. Filippov, 'Differencialnyje uravnenija s razryvnoj pravoj chastju' (in Russian), Nauka Publishing House, 1985.

[12] A. F. Filippov, 'Differential equations with discontinuous righthand sides', Kluwer Academic Publishers, 1988.

[13] T. Hanus, D. Janovská, 'Discontinuous, piecewise–smooth dynamical systems', CD-ROM of full texts CHISA 2006, Czech Society of Chemical Engineering, 2006.

[14] M. W. Hirsch, S. Smale, R. L. Devaney, 'Differential equations, dynamical systems and an introduction to chaos', Elsevier, Academic Press, 2003.

[15] M. C. Irwin, 'Smooth dynamical systems', Elsevier, Academic Press, 1980.

[16] D. Janovská, T. Hanus, M. Biák, 'Some applications of piece-wise smooth dynamical systems', ICNAAM 2010, AIP conference proceedings, 1281, American Institute of Physics, 2010, pp. 728–731.

[17] D. Janovská, T. Hanus, 'Qualitative methods in discontinuous dynamical systems', ICNAAM 2011, AIP conference proceedings, 1389, American Institute of Physics, 2011, pp. 1252–1255.

[18] D. Janovská, M. Biák, T. Hanus, 'Some applications of Filippov's dynamical systems', Journal of Computational and Applied Mathematics, 254, 2013, pp. 132–143.

[19] M. Kunze, 'Non-smooth dynamical systems', Springer-Verlag, 2000.

[20] Y. A. Kuznetsov, S. Rinaldi, A. Gragnani, 'One-parameter bifurcations in planar Filippov systems', International Journal of Bifurcation and Chaos, 2003, pp. 2157–2188.

[21] Y. A. Kuznetsov, 'Elements of applied bifurcation theory', 1st. ed., Springer-Verlag, 1995.

[22] R. I. Leine, H. Nijmeijer, 'Dynamics and bifurcations of non-smooth mechanical systems', Springer-Verlag, 2004.

[23] E. Mosekilde, Z. Zhusubalyev, 'Bifurcations and chaos in piecewise-smooth dynamical systems', World Scientific Publishing, 2003.

[24] K. Popp, N. Hinrichs, M. Oestreich, 'Dynamical behaviour of friction oscillators with simultaneous self and external excitation', Sadhana, Indian Academy of Sciences, 20, 1995, pp. 627–654.

[25] K. Popp, P. Shelter, 'Stick-slip vibrations and chaos', Philosofical Transactions of the Royal Society A, 332 (1624), 1990, pp. 89–105.
[26] C. Robinson, 'Dynamical systems : stability, symbolic dynamics, and chaos', CRC Press, 1998.
[27] M. Wiercigroch, B. de Kraker, 'Applied nonlinear dynamics and chaos of mechanical systems with discontinuities', World Scientific Publishing, 2000.

6

Numerical Methods and Some of Their Applications

A. Munjal and J. Kaur

Department of Mathematic, Akal University, Talwandi Sabo, Bathinda 151302, Punjab, India

Abstract

This chapter is primarily on the understanding of the numerical techniques. Numerical techniques are used to approximate the solution of differential equations which cannot be solved analytically. It investigates the history and the factors that must be considered to establish the new numerical method. It deals with the types of numerical methods. It includes distinct numerical methods with their order of convergence. This chapter describes the numerical analysis theory for system of linear equations and for system of non-linear equations. It explains how and why components of engine degrade. The domain of applications of numerical methods includes a broad range of engineering, mathematical physics, and linear algebra, which are described in this chapter.

Keywords: Differential equations, buffet, Lipschitz-condition, continuous function, Numerical techniques, Order of convergence.

Numerical techniques are the methods used to approximate the solution of differential equations numerically. There are many differential equations which cannot be solved analytically; so these methods are very useful to find out the solution of those problems. In engineering and daily life, approximation to the solution is sufficient. So, numerical techniques are required to approximate the solution. Jayakumar *et al.* [4] solved hybrid fuzzy differential equations with the help of Runge–Kutta Heun numerical method. Runge–Kutta Nyström methods adapted to the numerical integration of perturbed oscillators are obtained. Franco [1] worked on the algorithms which integrate exactly harmonic oscillators. He described the necessary and sufficient conditions for the required order of Nyström numerical method. Differential Riccati equations play a fundamental role in control theory, for example, optimal theory, decoupling, order reduction, etc. Peinado and Hernaìndez [5] used Adam-Bashforth numerical methods to solve differential Riccati equations. Uncertain differential equations are driven by Liu process and have been widely applied to the uncertain finance. It is not possible to obtain the solution of these equations analytically. So, Rang Gao [2] used numerical methods to solve these equations such as Milne method. Hager [3] used optimal Runge–Kutta method in optimal control and the transformed adjoin system.

6.1 History of Numerical Analysis

Egyptian Rhind papyrus (1650 BC) described root finding method for solving simple equation. After this, Greek mathematician worked in this area and made further advancements. Eudoxus (400–350 BC) of Cnidus established the method of exhaustion for calculating length, area, and volume of geometric figure. Archimedes (285–212 BC) modified this method. This method played an important role in development of calculus which led to accurate mathematical models. These models were too complicated to solve analytically; so efforts were made to obtain approximate solution for these models. This type of solutions gave rise to numerical analysis.

In 1614, Scottish mathematician John Napier created logarithms for tedious calculations. These logarithms changed multiplication and division to simple addition and subtraction. Newton (1642–1727) (1707–1783) created a number of numerical methods for solving a variety of problems. Swiss mathematician Leonhard Euler (1707–1783) and French mathematician Joseph-Louis Lagrange (1736–1813) worked in this region. Leonhard Euler established a numerical method named "Euler method" and Lagrange proved the convergence of this method. Euler method is the simplest and the most analyzed numerical technique. Abraham worked on the improvement of Euler method and named it modified Euler method and its order of convergence increased to 2. German mathematician Carl Friedrich Gauss (1777–1855) made a major contribution in numerical analysis. He invented the "Gaussian elimination method." This can also be used to find the rank and determinant of the matrix and inverse of invertible matrix. Numerical analysis is the basic tool for mathematical model in science given by Newton to describe the effect of gravity.

German mathematicians Carl Runge (1856–1927) and Martin Kutta (1867–1944) developed the Runge–Kutta method which is a great invention in numerical analysis. These methods are widely used by physicists to numerically solve the differential equation. Runge–Kutta methods of order 2 and order 4 are famous numerical methods to solve the initial value problems.

6.2 Theory of Numerical Analysis

The implementation of a numerical method with an appropriate convergence check in programming language is called a numerical algorithm. The main idea in numerical methods is that we convert the differential equation into the difference equation, and then solve the problem. There are methods which convert the partial differential equations to the ordinary differential equations (ODEs). We can even convert the higher order differential equation into the first-order differential equation.

For example, if we have a differential equation of an order m as

$$\begin{aligned} y^m(t) &= F(t,\, y(t),\, y'(t), \ldots, y^{m-1}(t)) \\ y^p(t_0) &= y_0^p \quad \text{for } p = 0,\ 1,\ 2, \ldots, m-1. \end{aligned} \tag{6.1}$$

To convert this equation to first-order differential equation, suppose

$$\begin{aligned} y &= v_1, \\ y' &= v_2, \\ y'' &= v_3, \\ &\ldots, \\ y^{m-1} &= v_m, \\ y^m &= F(t,\, v_1, v_2, \ldots,\, v_m). \end{aligned}$$

Then, we get

$$\frac{dv}{dt} = F(t,\, v) \quad v(t_0) = \eta_0 \tag{6.2}$$

where $v = [v_1, v_2, \ldots, v_m]^t$ and $F(t,\, v) = [v_1, v_2, \ldots,\, v_m,\, F(t,\, v_1, v_2, \ldots, v_m)]^t$ and $\eta_0 = [y_0,\, y_0', \ldots, y_0^m]^t$.

Hence, *m* order differential equation is converted into first-order differential equation. So, without loss of generality of higher order system, we just deal with the first-order differential equations. However, before attempting to approximate the solution numerically, we must check that whether the differential equation has solution or not. This can be answered for the ODEs by the following theorem.

6.2.1 Theorem

Let *D* be an open connected set in $\mathbb{R}^2$ and let $f(t,\ y)$ be a continuous function of t and y, where $(t,\ y) \in D$. Let $(t_0,\ y_0)$ be the interior point of *D*. Assume that*f* satisfies the Lipschitz-condition; then there exists a unique solution of the differential equation

$$\begin{aligned} Y^{'}(t) &= f(t,\ Y(t)) \\ Y(t_0) &= y_0 \end{aligned} \tag{6.3}$$

on the interval $[t-h,\ t+h]$ for some $h > 0$.

Note: Here, continuity of the function $f(t,\ y)$ guarantees the existence of the solution of the differential equation and Lipschitz condition guarantees the existence of the unique solution.

Notation: Let us consider the differential equation

$$\begin{aligned} Y^{'}(t) &= f(t,\ Y(t)) \\ Y(t_0) &= y_0. \end{aligned} \tag{6.4}$$

Let $Y_0(t)$ denote the true solution of the above differential equation and a continuous function for $a \leq t \leq b$. Let y_i denote the value of true solution at $t = t_i$ and w_i denote the approximated solution of the above differential equation obtained at $a < t_1 < t_2 < \cdots < t_{N-1} < t_N = b$.

6.3 Characterization of Numerical Methods

6.3.1 One-Step or Single-Step Method

The methods in which computation of w_{i+1} requires only the knowledge of w_i, where w_i is the approximation, made one step prior to the present value.

General form of linear one-step method is

$$\frac{w_{i+1} - w_i}{h_i} - \phi(f,\ t_i,\ w_i,\ w_{i+1},\ h_i) \tag{6.5}$$

where $h_i = t_{i+1} - t_i$.

6.3.2 Multi-Step Method

The methods in which computation of w_{i+1} does not require only the knowledge of w_i, where w_i is the approximation, made one step prior to the present value but also some more values prior to w_i.

General form of linear multi-step method

$$\frac{w_{i+1} - a_1 w_2 - \cdots - a_m w_{i-(m-1)}}{h_i}$$
$$= b_0 f(t_{i+1},\ w_{i+1}) + b_1 f(t_i,\ w_i) + \cdots + b_m f(t_{i-(m-1)}, w_{i-(m-1)}) \quad (6.6)$$

where $h_i = t_{i+1} - t_i$.

Each numerical method takes an initial point as first approximation and, with the help of initial point, goes to the next step to find the next solution point. The process will go on to find out the solution. Single-step methods use only one previous point and its derivative to calculate the next approximation and discard all previous information, but multi-step methods use the information from previous steps rather than discarding it. So, we can say that multi-step numerical methods are more efficient than the single-step numerical methods.

6.4 Types of Numerical Methods

6.4.1 Explicit Method

If the function ϕ in Equation (6.5) is independent of w_{i+1}, then the difference equation can be solved explicitly for w_{i+1}. So, the method is said to be an explicit method, and if $b_0 = 0$ in Equation (6.6), then the multi-step method becomes explicit. These methods are made accurate easily as compared to the implicit method as explicit methods produce greater accuracy with less computational effort than implicit methods. Non-stiff problems are solved by using explicit methods because we can achieve acceptable accuracy with minimal costs. FLOW-3D uses explicit techniques in normal cases. Euler method is an example of an explicit method.

6.4.2 Implicit Method

If the function θ in Equation (6.5) is dependent on w_{i+1}, then the difference equation can be solved implicitly for w_{i+1}. So, the method is said to be an implicit method, and if $b_0 \neq 0$ in Equation (6.6), then the multi-step method becomes implicit. These methods are more stable as compared to the explicit method. Stiff problems are solved by implicit methods. In computational fluid dynamics, we use the implicit method. Backward Euler method is an example of an explicit method.

6.5 Factors

The factors that must be considered to establish the new numerical methods are as follows:

1) consistency of method;
2) convergence of method;
3) stability of method.

6.5.1 Consistency of Method

To establish any numerical method, the first factor that we have to consider is the consistency; if the method is consistent, then it is the good approximation of the differential equation. The error associated with the approximation of the differential equation is called the local truncation error which we will denote by a symbol τ_i. It is called a local error because it measures the error generated by the method, assuming the solution at previous steps was exact. This error measures how well the difference equation that defines the method approximates the differential equation.

For one-step method:

$$\tau_i = \frac{y_{i+1} - y_i}{h_i} - \phi(f,\, t_i,\, y_i,\, y_{i+1},\, h_i). \tag{6.7}$$

For multi-step method:

$$\tau_i = \frac{y_{i+1} - \sum\limits_{j=1}^{m} a_j y_{i+1-j}}{h_i} - \sum_{j=0}^{m} b_j f(t_{i+1-j},\, y_{i+1-j}). \tag{6.8}$$

If $\tau_i \to 0$ as $h_i \to 0$, then the difference equation is said to be consistent with the differential equation. Thus, the numerical method is consistent in nature.

6.5.2 Convergence of Method

If the numerical method is consistent, then check whether the method is convergent or not. It is one of the main properties of the numerical method.

The error associated with the difference between the solution of differential equation, say y_i, and the solution of difference equation, say w_i, i.e., $y_i - w_i$, is called global discretization error. We refer to this error as global error because it measures the cumulative effect errors that occurred at all steps in the method.

If $\lim_{h_i \to 0} \max\{|y_i - w_i| : 1 < i < N\} = 0$, then we say that the numerical method is convergent.

6.5.3 Stability of Method

The numerical method is said to be stable if small perturbations in initial value causes a small change in numerical solution. We can also measure the stability of the method z. Let $[a < t_1 < t_2 < \cdots < t_{N-1} < t_N = b]$ be the partition of $[a,\ b]$ for the numerical method. Let $\{w_i\}$ denote the sequence of approximated solution of the differential equation from the initial condition α and $\{\overline{w}_i\}$ denote the sequence of approximated solution of the differential equation from the initial condition $\overline{\alpha}$. Then the numerical method is said to be stable iff there exists a function $k(t) > 0 \ni |\overline{w}_i - w_i| \leq k(t_i)|\alpha - \overline{\alpha}|$.

6.6 Distinct Numerical Methods

6.6.1 Euler Method

Swiss mathematician Leonhard Euler who is one of the most eminent and successful mathematicians in the history of the mathematics. In the beginning of the 18th century, Euler described the polygonal method to solve many systematic, and difficult problems arose which could be not solved by basic methods formulated by Newtonian mechanics. This polygonal method created the numerical method, and because of its applications, history has remembered it as Euler's method. It is an explicit numerical method for solving differential equations. Euler's method solves certain kinds of first-order differential equations in CE system. It is helpful in estimating the force-deformation behavior as it is very easy to apply. Consider the differential Equation (6.1) and let $[a = t_0 < t_1 < t_2 < \cdots < t_{N-1} < t_N = b]$ be the partition of $[a,\ b]$ with mesh points being equally spaced, i.e., $t_n = t_0 + nh$ for $n = 0, 1, 2, \ldots, N$.

Take $w_0 = y_0$

Then the approximated solution by Euler method is given by

$$w_{n+1} = w_n + hf(t_n, w_n). \tag{6.9}$$

Order of Euler method: To obtain the order of Euler method, we will examine the local truncation error τ_i of Euler method.

From the definition of Euler Method, it is given by

$$\tau_i = \frac{y_{i+1} - y_i}{h_i} - f(t_i,\, y_i). \tag{6.10}$$

By Taylor series expansion,

$$y(t_{i+1}) = y(t_i) + hy^{'}(t_i) + \frac{h^2}{2!}y^{''}(\xi_i) \text{ where } t_i < \xi_i < t_{i+1}$$

$$\Rightarrow y_{i+1} = y_i + hf(t_i,\, y_i) + \frac{h^2}{2!}y^{''}(\xi_i) \text{ where } t_i < \xi_i < t_{i+1}.$$

Therefore,

$$\tau_i = \frac{h}{2!}y^{''}(\xi_i) \text{ where } t_i < \xi_i < t_{i+1}.$$

$$\Rightarrow |\tau_i| \leq \frac{h}{2}\max\{|y^{''}(t)| : a \leq t \leq b\}. \tag{6.11}$$

Clearly, τ_i goes like the first power of step size. Hence, it is a first-order scheme.

Also, as $h_i \to 0, \tau_i \to 0$. So this implies that Euler method is consistent.

6.6.2 Backward Euler Method

This is the extension of Euler method. It is an implicit method. In this method, we use backward difference approximation instead of forward difference approximation. Consider the differential Equation (6.1) and let $[a = t_0 < t_1 < t_2 < \cdots < t_{N-1} < t_N = b]$ be the partition of $[a,\, b]$ with mesh points being equally spaced, i.e., $t_n = t_0 + nh$ for $n = 0, 1, 2, \ldots, N$.

Take $w_0 = y_0$.

Then the approximated solution by Euler method is given by

$$w_{n+1} = w_n + hf(t_{n+1}, w_{n+1}). \tag{6.12}$$

To implement backward Euler method, we use the predictor formula which is given by

$$w_{n+1} = w_n + hf(t_{n+1}, w_n + hf(t_{n+1},\, y_n)). \tag{6.13}$$

This is also of order one. Hence, it is a consistent method.

6.6.3 Runge–Kutta Methods

In 1895, German mathematicians Carl Runge and Wilhelm Kutta generalized the approximation method of Euler to more appropriate numerical methods of better accuracy. The idea of Euler was to propagate the solution of an initial value problem forward by a sequence of small time-steps. It was adopted for Runge–Kutta methods also. These methods include both implicit and explicit iterative methods. In modern software for the solution of initial value, problems are mainly derived with the help of these methods. As we know that it is better to solve non-stiff problems by using explicit methods, we can achieve acceptable accuracy with minimal costs. However, for stiff problems, stability rather than accuracy becomes the dominant consideration; hence, implicit methods become the more appropriate choice in that case. New Zealand mathematician John Charles Butcher works on multi-step methods for initial value problems such as Runge–Kutta methods. The Butcher group and the Butcher tableau are named after him.

6.6.3.1 Runge–Kutta method of an order 2

These are used in temporal discretion for the approximate solutions of ODEs.

We will begin with Runge–Kutta method of order 2, and, later on, we consider some higher order

$$y_{n+1} = y_n + hF(t_n,\ y_n; h) \ \ n \geq 0. \tag{6.14}$$

For methods of order 2, we generally choose

$$F(t,\ y; h) = b_1 f(t, y) + b_2 f(t + \alpha h,\ y + \beta h f(t,\ y) \tag{6.15}$$

where $\alpha,\ \beta,\ b_1,$ and b_2 are constants and we have to determine their values such that truncation error

$$\tau_n = \frac{y_{n+1} - y_n}{h} - F(t_n,\ y_n;) \tag{6.16}$$

will satisfy $\tau_n = O(h^2)$. After solving this value by using taylor series expansion on $F(t,\ y; h)$, we get

$$b_1 + b_2 = 1$$

and

$$\alpha = \frac{1}{2b_2} = \beta.$$

There are three types of Runge–Kutta methods of an order 2 depending on the value of b_2.

6.6.3.1.1 Heun's method

German mathematician Karl Heun made significant contributions for development of the Runge–Kutta method. In 1900, this is also known as improved Euler method. Runge–Kutta method is termed as Heun's method if we take $b_2 = \frac{1}{2}$. So, $b_1 = \frac{1}{2}$ and $\alpha = \beta = 1$.

Therefore,

$$y_{n+1} = y_n + \frac{h}{2}[f(t_n, y_n) + f(t_n + h, y_n + hf(t_n, y_n)). \tag{6.17}$$

6.6.3.1.2 Modified Euler method

This is an improved Euler method. This method is widely used in medicine, biology, and anthropology. Runge–Kutta method is termed as modified Euler method; if we take $b_2 = 1$. So, $b_1 = 0$ and $\alpha = \beta = \frac{1}{2}$.

Therefore,

$$y_{n+1} = y_n + h\left[f(t_n + \frac{h}{2}, y_n + \frac{h}{2}f(t_n, y_n)\right]. \tag{6.18}$$

6.6.3.1.3 Optimal Runge–Kutta method

ODEs arise frequently in the study of physical systems; this method is very useful in that cases. Runge–Kutta method is termed as optimal Runge–Kutta method, if $b_2 = \frac{3}{2}$, $b_1 = \frac{1}{4}$, $\alpha = \beta = \frac{2}{3}$.

Therefore,

$$y_{n+1} = y_n + \frac{h}{4}\left[f(t_n, y_n) + 3f(t_n + \frac{2}{3}h, y_n + \frac{2}{3}hf(t_n, y_n)\right]. \tag{6.19}$$

6.6.3.2 Runge–Kutta method of an order 3

$$z_1 = y_n$$
$$z_2 = y_n + \alpha_{2,1}f(t_n, z_1)$$
$$z_3 = y_n + \alpha_{3,1}hf(t_n, z_1) + \alpha_{3,2}hf(t_n + c_2h, z_2).$$

Then

$$y_{n+1} = y_n + h[b_1f(t_n, z_1) + b_2f(t_n + c_2h, z_2) + b_3f(t_n + c_3h, z_3)] \tag{6.20}$$

where $\alpha_{2,1}, \alpha_{3,1}, \alpha_{3,2}, c_2, c_3, b_1, b_2,$ andb_3 are constants and we determine their values such that truncation error

$$\tau_n = \frac{y_{n+1} - y_n}{h} - [b_1f(t_n, z_1) + b_2f(t_n + c_2h, z_2) + b_3f(t_n + c_3h, z_3)] \tag{6.21}$$

will satisfy $\tau_n = O(h^3)$. After solving this value by using taylor series expansion, we get

$$c_2 = \alpha_{2,1}$$

$$\alpha_{3,2} + \alpha_{3,1} = c_3$$

$$b_1 + b_2 + b_3 = 1$$

$$c_2 b_2 + c_3 b_3 = \frac{1}{2}$$

$c_2^2 b_2 + c_3^2 b_3 = \frac{1}{3}$ and

$$c_2 \alpha_{3,2} b_3 = \frac{1}{6}.$$

There are two types of Runge–Kutta methods of an order 3 depending on the values of these constants.

6.6.3.2.1 Nyström method

In 1924, Nyström described this method. Runge–Kutta method is termed as Nyström method, if we take $c_2 = \alpha_{2,1} = \frac{2}{3}$, $\alpha_{3,1} = 0$, $\alpha_{3,2} = \frac{2}{3}$, $c_3 = \frac{2}{3}$, $b_1 = \frac{2}{8}$, $b_2 = \frac{3}{8}$, $b_3 = \frac{3}{8}$.

Therefore,

$$y_{n+1} = y_n + \frac{h}{8}[2f(t_n, z_1) + 3f(t_n + \frac{2}{3}h, z_2) + 3f(t_n + \frac{2}{3}h, z_3)]. \quad (6.22)$$

6.6.3.2.2 Classical Runge–Kutta method

Despite the evolution of a vast and comprehensive body of knowledge, it continues to be a source of active research. Runge–Kutta method is termed as classical Runge–Kutta method, if we take $c_2 = \alpha_{2,1} = \frac{1}{2}$, $\alpha_{3,1} = -1$, $\alpha_{3,2} = 2$, $c_3 = 1$, $b_1 = \frac{1}{6}$, $b_2 = \frac{4}{6}$, $b_3 = \frac{1}{6}$.

Therefore,

$$y_{n+1} = y_n + \frac{h}{6}\left[f(t_n, z_1) + 4f(t_n + \frac{1}{2}h, z_2) + f(t_n + h, z_3)\right]. \quad (6.23)$$

6.6.3.3 Runge–Kutta method of an order 4

$$z_1 = y_n$$

$$z_2 = y_n + \alpha_{2,1} h f(t_n, z_1)$$

$$z_3 = y_n + \alpha_{3,1} h f(t_n, z_1) + \alpha_{3,2} h f(t_n + c_2 h, z_2)$$

$$z_4 = y_n + \alpha_{4,1}hf(t_n,\ z_1) + \alpha_{4,2}hf(t_n + c_2h,\ z_2) + \alpha_{4,3}hf(t_n + c_3h,\ z_3).$$

Then

$$y_{n+1} = y_n + h[b_1f(t_n, z_1) + b_2f(t_n + c_2h,\ z_2) + b_3f(t_n + c_3h,\ z_3) + b_4f(t_n + c_4h,\ z_4)] \tag{6.24}$$

where $\alpha_{2,1},\ \alpha_{3,1},\ \alpha_{3,2},\ \alpha_{4,1}, \alpha_{4,2},\ \alpha_{4,3},\ c_2,\ c_3,\ c_4,\ b_1,\ b_2,\ b_3, b_4$ are constants and we determine their values such that truncation error

$$\tau_n = \frac{y_{n+1} - y_n}{h} - [b_1f(t_n, z_1) + b_2f(t_n + c_2h,\ z_2) + b_3f(t_n + c_3h,\ z_3) + b_4f(t_n + c_4h,\ z_4)] \tag{6.25}$$

will satisfy $\tau_n = O(h^4)$. After solving this value by using taylor series expansion, we get

$$c_2 = \alpha_{2,1}$$

$$c_3 = \alpha_{3,1} + \alpha_{3.2}$$

$$c_4 = \alpha_{4,1} + \alpha_{4.2} + \alpha_{4,3}$$

$$b_1 + b_2 + b_3 + b_4 = 1$$

$$b_1c_2 + b_2c_3 + b_3c_4 = \frac{1}{2}b_1c_2^2 + b_2c_3^2 + b_3c_4^2 = \frac{1}{3}b_1c_2^3 + b_2c_3^3 + b_3c_4^3 = \frac{1}{4}$$

$$b_3c_2\alpha_{3,2} + b_4(c_2\alpha_{4,2} + c_3\alpha_{4,3}) = \frac{1}{6}$$

$$b_3c_3c_2\alpha_{3,2} + b_4c_4(c_2\alpha_{4,2} + c_3\alpha_{4,3}) = \frac{1}{8}$$

$$b_3c_2^2\alpha_{3,2} + b_4(c_2^2\alpha_{4,2} + c_3^2\alpha_{4,3}) = \frac{1}{6}$$

$$b_4c_2\alpha_{3,2}\alpha_{4,3} = \frac{1}{24}.$$

6.6.3.3.1 Classical Runge–Kutta method

In 1901, Kutta gave the classical fourth-order Runge–Kutta method. This is the most widely known member of the Runge–Kutta family is written as "RK4." It is the ODE analog of Simpson's method for numerical integration. Runge–Kutta method is termed as classical Runge–Kutta method, if we take $c_2 = \alpha_{2,1} = \frac{1}{2},\ \alpha_{3,1} = 0,\ \alpha_{3,2} = \frac{1}{2},\ \alpha_{4,1} = 0,\ \alpha_{4,2} = 0,\ \alpha_{4,3} = 1, c_3 = \frac{1}{2},\ c_4 = 1,\ b_1 = \frac{1}{6},\ b_2 = \frac{2}{6},\ b_3 = \frac{2}{6},\ b_4 = \frac{1}{6}.$

Therefore,

$$y_{n+1} = y_n + \frac{h}{6}\left[f(t_n,\ z_1) + 2f\left(t_n + \frac{1}{2}h,\ z_2\right) + 2f\left(t_n + \frac{1}{2}h,\ z_3\right) + 2f(t_n + h,\ z_4)\right]. \tag{6.26}$$

6.6.4 Adam's Bashforth Method

John Couch Adams to established the numerical method to solve a differential equation modeling capillary action due to Francis Bashforth. In 1833, Bashforth published his theory and Adam's numerical method known as Adam's Bashforth method. This is an example of multi-step explicit method. Since Adam's Bashforth method of an order p involves prior p terms, we are given only with one initial value; so we use taylor series to find the remaining terms upto $p\text{th}$ order at initial condition.

6.6.4.1 Adam's Bashforth method of an order 2

$$y_{n+1} = y_n + \frac{h}{2}[3f(t_n, y_n) + f(t_{n-1}, y_{n-1})]. \tag{6.27}$$

6.6.4.2 Adam's Bashforth method of an order 3

$$y_{n+1} = y_n + \frac{h}{12}[23f(t_n, y_n) - 4f(t_{n-1}, y_{n-1}) + 5f(t_{n-2}, y_{n-2})]. \tag{6.28}$$

6.6.4.3 Adam's Bashforth method of an order 4

$$y_{n+1} = y_n + \frac{h}{24}[55f(t_n, y_n) - 59f(t_{n-1}, y_{n-1}) + 37f(t_{n-2}, y_{n-2}) - 9f(t_{n-3}, y_{n-3})]. \tag{6.29}$$

6.6.5 Milne-Simpson's Method

In 1953, Milne described the predictor-corrector scheme for solving differential equations, which is known as the Milne or Milne-Simpson method. This is an example of implicit multi-step method. Since Milne-Simpson's method of an order p involves prior $p-1$ terms, but we are given only with one initial value; so we use taylor series to find the remaining terms upto pth order at initial condition.

6.6.5.1 Milne-Simpson's method of an order 2

$$y_{n+1} = y_n + \frac{h}{2}[f(t_{n+1}, y_{n+1}) + f(t_n, y_n)]. \tag{6.30}$$

6.6.5.2 Milne-Simpson's method of an order 3

$$y_{n+1} = y_n + \frac{h}{12}[5f(t_{n+1}, y_{n+1}) + 8f(t_n, y_n) - f(t_{n-1}, y_{n-1})]. \tag{6.31}$$

6.6.5.3 Milne-Simpson's method of an order 4

$$\begin{aligned} y_{n+1} = y_n + \frac{h}{24}[9f(t_{n+1}, y_{n+1}) + 19f(t_n, y_n) - 5f(t_{n-1}, y_{n-1}) \\ + f(t_{n-2}, y_{n-2})]. \end{aligned} \tag{6.32}$$

6.7 Numerical Solution in Aircraft Buffet

F-35 is one of the famous aircrafts in service. One of the problems encountered by the person who is flying it first is that it buffets. The definition of buffet given by the first pilot of F-35 is that it shakes when it goes at high geologies, i.e., shaking at high geologies is known as buffeting.

Buffeting is both useful and harmful. Advantage of buffeting is that it tells the pilot that plane is cooling at high angle of attack. Disadvantage is that it shakes too much that pilot have trouble to see the screen clearly or what is displayed on screen; it creates problem.

Initially buffet is very coherent, regular, and rose like stream, but, after burst, it becomes irregular and incoherent that goes right to tail which makes the aircraft shake. It is similar to the partial differential equation. Thus, we can receive the information about buffeting either from the plane test or using numerical methods for partial differential equation. The main difference is that if we want improvement in buffet, then there are two cases. One is to go with numerical solution for corresponding partial differential equation and the other is plane test. But if we go with plane test, then we need other planes to apply this which is not so easy practically. Hence, numerical methods are so important in the engineering.

6.8 Numerical Solution in Degradation of Components of Engine

Engine plays an important role in our daily life and numerical methods are very useful to know why the components of engine degrade. Inside the engine, air goes through many stages of compressor. Then, air passes to the pretty small combustion chamber that burns; after the combustion chamber, power should be extracted through the several rolls of blades. So, these blades create the main trouble for the person who is working on the design of engine. Since there is high temperature as much as we can make, also it is harder than melting temperature of any matter. Hence, we cannot put any camera inside the engine to see why the blades get damaged and the components of engine get degraded. So, we use the numerical solutions of partial differential equations to get rid of this problem because numerical simulation governs the flow.

6.9 Applications of Numerical Techniques

The domain of applications of numerical methods includes a broad range of engineering, mathematical physics, and linear algebra.

1. Numerical methods are used in computer science to define algorithm for finding roots of equations and for multi-dimensional root finding.
2. Engineers describe and predict the behavior of systems with the help of numerical methods.
3. Numerical methods play a vital role in train and traffic signals.
4. These are also used in electromagnetic analysis for detection by radar.
5. Numerical techniques play a vital role in research, and it is the most recent topic in research area.
6. Computational fluid dynamics (CFD) uses numerical techniques in weather prediction.
7. They have wide range of applications in network simulation.
8. Regional uptake of inhaled materials by respiratory tract uses numerical techniques.
9. There are some scientific programs which are completely based on numerical techniques.
10. Used in modeling combustion flow in coal power plant.
11. Mathematical problems can be solved with arithmetic operations.

References

[1] Franco J. M. (2002) Runge–Kutta Nyström methods adapted to the numerical integration of perturbed oscillators, Computer Physics Communications, 147(3), 770–787.

[2] Gao R. (2016) Milne method for solving uncertain differential equations, Applied Mathematics and Computation, 274, 774–785.

[3] Hager W. (2001) Runge–Kutta methods in optimal control and the transformed adjoint system, Numerische Mathematik, 87(2). 35–49.

[4] Jayakumar T., Kanagarajan K. and Sangameswaran I. (2014) Numerical solution for hybrid fuzzy systems by Runge–Kutta Heun method, Far East Journal of Mathematical Sciences, 2(2), 67–90.

[5] Peinado J. and Hernaìndez V. (2010) Adams-Bashforth and Adams-Moultan methods for solving differential Riccati equations, Computers and Mathematics with applications, 60(11), 3032–3045.

7

On a Certain Subclass of Analytic Functions Defined by q-Analogue Differential Operator

B. Venkateswarlu[1], P. Thirupathi Reddy[2], S. Sridevi[3] and Sujatha [4]

[1,3,4]Department of Mathematics, GSS, GITAM University,
Doddaballapur- 562 163, Bengaluru Rural, Karnataka, India.
[2]Department of Mathematics, Kakatiya University,
Warangal-506 009, Telangana, India.
E-mail: bvlmaths@gmail.com; siri_settipalli@yahoo.co.in;
sujathavaishnavy@gmail.com; reddypt2@gmail.com

Abstract

This chapter attempts to define q-analogue of Ruscheweyh differential operator and study some results related to coefficient estimates, neighborhoods, and partial sums related to the subclass.

Keywords and Phrases: analytic functions, coefficient bounds, partial sums.

AMS Subject Classification: 30C45, 30C80.

7.1 Introduction

Let A denote the class of all functions u of the form

$$u(z) = z + \sum_{\eta=2}^{\infty} a_\eta z^\eta \tag{7.1}$$

in the open unit disc $\Delta = \{z \in \mathbb{C} : |z| < 1\}$. The subclasses of A having univalent and normalization $(u(0) = u'(0) - 1 = 0)$, which was indicated by S.

Functions $u \in A$ and $z \in \Delta$ are star-like and convex functions of the order $\varsigma, 0 \leq \varsigma < 1$, if it satisfies the following, respectively:

$$\Re\left\{\frac{zu'(z)}{u(z)}\right\} > \varsigma \tag{7.2}$$

$$\Re\left\{1 + \frac{zu''(z)}{u'(z)}\right\} > \varsigma. \tag{7.3}$$

We denote these classes with $S^*(\varsigma)$ and $K(\varsigma)$, respectively.

Let T denote the class of analytic function in Δ that is of the form

$$u(z) = z - \sum_{\eta=2}^{\infty} a_\eta z^\eta, \quad (a_\eta \geq 0), \tag{7.4}$$

and this was studied by Silverman [12].

The convolution of two functions u and v is

$$(u * v)(z) = z + \sum_{\eta=2}^{\infty} a_\eta b_\eta z^\eta = (v * u)(z),$$

where

$$v(z) = z + \sum_{\eta=2}^{\infty} b_\eta z^\eta.$$

In [10], Sakaguchi introduced the class ST_s of star-shaped mappings w.r.t. symmetric points as follows:

$$\Re\left\{\frac{2zu'(z)}{u(z) - u(-z)}\right\} > 0, \quad (z \in \Delta),$$

and Owa *et al.* [6] introduced the class $ST_s(v, \varsigma)$ as follows:

$$\Re\left\{\frac{(1-\varsigma)zu'(z)}{u(z) - u(\varsigma z)}\right\} > v, \quad (0 \leq v < 1, |\varsigma| \leq 1, \varsigma \neq 1, z \in \Delta).$$

Note that $ST_s(0, -1) = ST_s$. Further, $ST_s(v, -1) = ST_s(v)$ is called Sakaguchi functions of order v.

Next, for $0 \leq \nu < 1$ and $k \geq 0$, the classes of $k-$uniformly starlike functions of order ν and convex functions of order ν, $k - UST(\nu)$ and $k - UCV(\nu)$ are, respectively, defined as follows:

$$\Re\left\{\frac{zu'(z)}{u(z)}\right\} > k\left|\frac{zu'(z)}{u(z)} - 1\right| + \nu,$$
$$\Re\left\{1 + \frac{zu''(z)}{u'(z)}\right\} > k\left|\frac{zu''(z)}{u'(z)}\right| + \nu, \quad (0 \leq \nu < 1,\ k \geq 0)$$

(for details, see [3]).

"The quantum calculus or $q-$calculus started with Frank Hilton Jackson in the early 20th century, but Euler and Jacobi had already worked out this kind of calculus. Recently, it is stimulated due to the vast requirement for mathematics that models quantum computing $q-$calculus, which seemed as an association between mathematics and physics. It requires plenty of application in different mathematical areas such as essential hypergeometric functions, quantum theory, mechanics, and the theory of relativity."

Definition 7.1. *For $0 < q < 1$, define the $q-$number $[\alpha]_q$ by*

$$[\alpha]_q = \begin{cases} \frac{1-q^\alpha}{1-q}, & \text{if} \quad \alpha \in \mathbb{C}, \\ \sum\limits_{i=0}^{\alpha-1} q = 1 + q + q^2 + \cdots + q^{n-1}, & \text{if} \quad \alpha = n \in \mathbb{N}. \end{cases} \tag{7.5}$$

Note that as $q \to 1^-$, $[n]_q \to n$.

Definition 7.2. *For $0 < q < 1$, define the $q-$fractional $[n]_q!$ by*

$$[n]_q! = \begin{cases} 1, & \text{if} \quad n = 0, \\ \prod\limits_{\eta=1}^{n} [\eta]_q, & \text{if} \quad n \in \mathbb{N}. \end{cases} \tag{7.6}$$

Definition 7.3. *Define the $q-$derivative $D_q u$ of a function u by*

$$(D_q)u(z) = \begin{cases} \frac{u(z)-u(zq)}{(1-q)z}, & \text{if} \quad z \neq 0, \\ u'(0), & \text{if} \quad z = 0 \end{cases} \tag{7.7}$$

provided that $u'(0)$ exists.

It follows from Equation (7.7) that

$$\lim_{q\to 1^-} D_q u(z) = \lim_{q\to 1^-} \frac{u(z)-u(zq)}{(1-q)z} = u'(z)$$

for a function u which is differentiable in a given subset of $\mathbb{C}$.
Thus, we have

$$(D_q)u(z) = 1 + \sum_{\eta=2}^{\infty} [\eta]_q a_\eta z^{\eta-1}. \tag{7.8}$$

Making use of $q-$operator, we generalize Salagean and Ruscheweyh differential operators as adopted.

Definition 7.4. *For $u \in A$, $\lambda \in \mathbb{N}_0 = \mathbb{N} \cup \{0\}$, the $q-$analogue of Salagean differential operator $\mathscr{I}_q^\lambda : A \to A$ is determined by*

$$\begin{aligned} \mathscr{I}_q^0 u(z) =& u(z) \\ \mathscr{I}_q^1 u(z) =& z\left(D_q u(z)\right) \\ &\vdots \\ \mathscr{I}_q^\lambda u(z) =& \mathscr{I}_q^1\left(\mathscr{I}_q^{\lambda-1} u(z)\right) = z\left(D_q \mathscr{I}_q^{\lambda-1} u(z)\right). \end{aligned}$$

Thus, we have

$$\mathscr{I}_q^\lambda u(z) = z + \sum_{\eta=2}^{\infty} [\eta]_q^\lambda a_\eta z^\eta. \tag{7.9}$$

Definition 7.5. *Let $u \in A$. Indicate by $\mathscr{R}_q^\lambda$, the $q-$analogue of Ruscheweyh differential operator defined by*

$$\mathscr{R}_q^\lambda u(z) = z + \sum_{\eta=2}^{\infty} \phi_\eta(\lambda, q) a_\eta z^\eta, \tag{7.10}$$

where $\phi_\eta(\lambda, q) = \frac{[\eta+\lambda-1]_q!}{[\lambda]_q![\eta-1]_q!}$ and $[\eta]_q!$ is defined as (7.6).

It may be remarked that when $q \to 1^-$, we have

$$\begin{aligned} \lim_{q\to 1^-} \mathscr{R}_q^\lambda u(z) =& z + \lim_{q\to 1^-} \sum_{\eta=2}^{\infty} \frac{[\eta+\lambda-1]_q!}{[\lambda]_q![\eta-1]_q!} a_\eta z^\eta \\ =& z + \sum_{\eta=2}^{\infty} \frac{(\eta+\lambda-1)!}{\lambda!(\eta-1)!} a_\eta z^\eta = \mathscr{R}^\lambda, \end{aligned}$$

where $\mathscr{R}^\lambda$ is Ruscheweyh differential operator that is given in [8] and has been examined by several researchers (for details, see [4, 5, 11]).

Now, we invent the definition of the class $k - UST_s(\lambda, q, \nu, \varsigma)$ by using the differential operator $\mathscr{R}_q^\lambda u(z)$.

Definition 7.6. *A function $u \in A$ is said to be in the class*

$$k - UST_s(\lambda, q, \nu, \varsigma), \quad (\lambda > 0, 0 < q < 1, 0 \leq \nu < 1, k \geq 0,$$

$$|\varsigma| \leq 1, \varsigma \neq 1, z \in \Delta)$$

if the adopting connection holds true:

$$\Re\left\{\frac{(1-\varsigma)z\left(\mathscr{R}_q^\lambda u(z)\right)'}{\mathscr{R}_q^\lambda u(z) - \mathscr{R}_q^\lambda u(\varsigma z)}\right\} \geq k\left|\frac{(1-\varsigma)z\left(\mathscr{R}_q^\lambda u(z)\right)'}{\mathscr{R}_q^\lambda u(z) - \mathscr{R}_q^\lambda u(\varsigma z)} - 1\right| + \nu.$$

Moreover, a function $u \in k - UST_s(\lambda, q, \nu, \varsigma)$ is in the subclass $k - \widetilde{U}ST_s(\lambda, q, \nu, \varsigma)$ if $u \in T$.

First, we need the adopting lemmas [1].

Lemma 7.1. *Let a be a complex number. Then*

$$\Re(a) \geq \alpha \Leftrightarrow \quad |a - (1+\alpha)| \leq |a + (1-\alpha)|.$$

Lemma 7.2. *Let a be a complex number and α, ν be real numbers. Then*

$$\Re\,(a) > \alpha|a-1| + \nu \quad \Leftrightarrow \quad \Re\{a(1+\alpha e^{i\varrho}) - \alpha e^{i\varrho}\} > \nu, \quad -\pi < \varrho < \pi.$$

The purpose of this chapter is to obtain coefficient bounds, partial sums, and certain neighborhood properties of class $k - \widetilde{U}ST_s(\lambda, q, \nu, \varsigma)$.

7.2 Coefficient Bounds

Theorem 7.1. *Let $u \in T$. Then $u \in k - \widetilde{U}ST_s(\lambda, q, \nu, \varsigma)$*

$$\Leftrightarrow \sum_{\eta=2}^{\infty} \phi_\eta(\lambda, q)\,|\eta(k+1) - u_\eta(k+\nu)|\,a_\eta \leq 1 - \nu, \tag{7.11}$$

where $u_\eta = 1 + \varsigma + \cdots + \varsigma^{\eta-1}$.
The estimate is sharp with

$$u(z) = z - \frac{1-\nu}{\phi_\eta(\lambda, q)\,|\eta(k+1) - u_\eta(k+\nu)|} z^\eta.$$

Proof. From Definition (7.1), we obtain

$$\Re\left\{\frac{(1-\varsigma)z\left(\mathscr{R}_q^{\lambda}u(z)\right)'}{\mathscr{R}_q^{\lambda}u(z)-\mathscr{R}_q^{\lambda}u(\varsigma z)}\right\}\geq k\left|\frac{(1-\varsigma)z\left(\mathscr{R}_q^{\lambda}u(z)\right)'}{\mathscr{R}_q^{\lambda}u(z)-\mathscr{R}_q^{\lambda}u(\varsigma z)}-1\right|+\nu.$$

Next, by Lemma 7.2, we have

$$\Re\left\{\frac{(1-\varsigma)z\left(\mathscr{R}_q^{\lambda}u(z)\right)'}{\mathscr{R}_q^{\lambda}u(z)-\mathscr{R}_q^{\lambda}u(\varsigma z)}(1+ke^{i\varrho})-ke^{i\varrho}\right\}\geq\nu,\qquad -\pi<\varrho<\pi$$

$$\Rightarrow\Re\left\{\frac{(1-\varsigma)z\left(\mathscr{R}_q^{\lambda}u(z)\right)'(1+ke^{i\varrho})}{\mathscr{R}_q^{\lambda}u(z)-\mathscr{R}_q^{\lambda}u(\varsigma z)}-\frac{ke^{i\varrho}\left[\mathscr{R}_q^{\lambda}u(z)-\mathscr{R}_q^{\lambda}u(\varsigma z)\right]}{\mathscr{R}_q^{\lambda}u(z)-\mathscr{R}_q^{\lambda}u(\varsigma z)}\right\}\geq\nu. \tag{7.12}$$

Now, suppose that

$$L(z)=(1-\varsigma)z\left(\mathscr{R}_q^{\lambda}u(z)\right)'(1+ke^{i\varrho})-ke^{i\varrho}\left[\mathscr{R}_q^{\lambda}u(z)-\mathscr{R}_q^{\lambda}u(\varsigma z)\right]$$

and

$$M(z)=\mathscr{R}_q^{\lambda}u(z)-\mathscr{R}_q^{\lambda}u(\varsigma z).$$

By virtue of Lemma 7.1, Equation (7.12) implies

$$|L(z)+(1-\nu)M(z)|\geq|L(z)-(1+\nu)M(z)|,\quad(0\leq\nu<1).$$

Hence, we obtain

$$\begin{aligned}
&|L(z)+(1-\nu)M(z)|\\
&\quad=\left|(1-\varsigma)\left\{(2-\nu)z-\sum_{\eta=2}^{\infty}\phi_\eta(\lambda,q)(\eta+u_\eta(1-\nu))a_\eta z^\eta\right.\right.\\
&\quad\left.\left.-ke^{i\varrho}\sum_{\eta=2}^{\infty}\phi_\eta(\lambda,q)(\eta-u_\eta)a_\eta z^\eta\right\}\right|\\
&\quad\geq|1-\varsigma|\left\{(2-\nu)|z|-\sum_{\eta=2}^{\infty}\phi_\eta(\lambda,q)\left|\eta+u_\eta(1-\nu)\right|a_\eta|z|^\eta\right.\\
&\quad\left.-k\sum_{\eta=2}^{\infty}\phi_\eta(\lambda,q)\left|\eta-u_\eta\right|a_\eta|z|^\eta\right\}.
\end{aligned}$$

On the other side, we get

$$
\begin{aligned}
&|L(z)+(1+\nu)M(z)| \\
&= \left|(1-\varsigma)\left\{-\nu z-\sum_{\eta=2}^{\infty}\phi_\eta(\lambda,q)(\eta+u_\eta(1-\nu))a_\eta z^\eta\right.\right. \\
&\quad \left.\left.-ke^{i\varrho}\sum_{\eta=2}^{\infty}\phi_\eta(\lambda,q)(\eta-u_\eta)a_\eta z^\eta\right\}\right| \\
&\quad \geq |1-\varsigma|\left\{\nu|z|-\sum_{\eta=2}^{\infty}\phi_\eta(\lambda,q)\,|\eta+u_\eta(1-\nu)|\,a_\eta\right. \\
&\quad \left.-k\sum_{\eta=2}^{\infty}\phi_\eta(\lambda,q)\,|\eta-u_\eta|\,a_\eta\,|z|^\eta\right\}.
\end{aligned}
$$

Accordingly, we find that

$$
\begin{aligned}
&|L(z)+(1-\nu)M(z)|-|L(z)+(1+\nu)M(z)| \\
&\quad \geq |1-\varsigma|\left\{2(1-\nu)|z|-\sum_{\eta=2}^{\infty}\phi_\eta(\lambda,q)\,[|\eta+u_\eta(1-\nu)|\right. \\
&\quad +|\eta-u_\eta(1+\nu)|+2k|\,|\eta-u_\eta|\,a_\eta\,|z|^\eta]\} \\
&\quad \geq 2(1-\nu)|z|-\sum_{\eta=2}^{\infty}2\phi_\eta(\lambda,q)\,|\eta(k+1)-u_\eta(k+\nu)|\,a_\eta\,|z|^\eta \geq 0
\end{aligned}
$$

or

$$
\sum_{\eta=2}^{\infty}\phi_\eta(\lambda,q)\,|\eta(k+1)-u_\eta(k+\nu)|\,a_\eta \leq 1-\nu.
$$

Conversely, suppose Equation (7.11) holds. Then, we must indicate that

$$
\Re\left\{\frac{(1-\varsigma)z\left(\mathscr{R}_q^\lambda u(z)\right)'(1+ke^{i\varrho})-ke^{i\varrho}\left[\mathscr{R}_q^\lambda u(z)-\mathscr{R}_q^\lambda u(\varsigma z)\right]}{\mathscr{R}_q^\lambda u(z)-\mathscr{R}_q^\lambda u(\varsigma z)}\right\} \geq \nu.
$$

Taking the values of $z(0 \leq |z| = r < 1)$ on the positive x-axis, then

$$\Re\left\{\frac{(1-\nu) - \sum_{\eta=2}^{\infty} \phi_\eta(\lambda, q)[\eta(1+ke^{i\varrho}) - u_\eta(\nu + ke^{i\varrho})]a_\eta z^{\eta-1}}{1 - \sum_{\eta=2}^{\infty} \phi_\eta(\lambda, q) u_\eta a_\eta z^{\eta-1}}\right\} \geq 0.$$

Since $\Re(-e^{i\varrho}) \geq -|e^{i\varrho}| = -1$, then

$$\Re\left\{\frac{(1-\nu) - \sum_{\eta=2}^{\infty} \phi_\eta(\lambda, q)[\eta(1+k) - u_\eta(\nu + k]a_\eta r^{\eta-1}}{1 - \sum_{\eta=2}^{\infty} \phi_\eta(\lambda, q) u_\eta a_\eta r^{\eta-1}}\right\} \geq 0.$$

If we let $r \to 1^-$, we get the result. □

Corollary 7.1. *If $u \in k - \widetilde{U}ST_s(\lambda, q, \nu, \varsigma)$, then*

$$a_\eta \leq \frac{1-\nu}{\phi_\eta(\lambda, q)\, |\eta(k+1) - u_\eta(k+\nu)|},$$

where $u_\eta = 1 + \varsigma + \cdots + \varsigma^{\eta-1}$.

7.3 Neighborhood Properties

In this section, using the concept of neighborhoods of analytic functions motivated by Goodman [2], Ruscheweyh [7], and Santosh [9], we define the neighborhood of a function $u \in T$.

Definition 7.7. *Let $\lambda > 0, 0 < q < 1, k \geq 0, |\varsigma| \leq 1, \varsigma \neq 1, 0 \leq \nu < 1, \upsilon \geq 0$ and $u_\eta = 1 + \varsigma + \cdots + \varsigma^{\eta-1}$. We define the $\upsilon-$neighborhood of a function $u \in T$ and indicated by $N_\upsilon(u)$ consisting of all functions $g(z) = z - \sum_{\eta=2}^{\infty} b_\eta z^\eta \in S$ $(b_\eta \geq 0)$ satisfying*

$$\sum_{\eta=2}^{\infty} \frac{\phi_\eta(\lambda, q)\, |\eta(k+1) - u_\eta(k+\nu)|}{1-\nu} |a_\eta - b_\eta| \leq 1 - \upsilon.$$

Theorem 7.2. *Suppose that* $u \in k - \widetilde{U}ST_s(\lambda, q, \nu, \varsigma)$ *and* $\Re(\nu) \neq 1$. *For any complex number* ϵ *with* $|\epsilon| < \upsilon (\upsilon \geq 0)$, *if* u *satisfies the pursuing stipulation:*

$$\frac{f(z) + \epsilon z}{1 + \epsilon} \in k - \widetilde{U}ST_s(\lambda, q, \nu, \varsigma)$$

then $N_\upsilon(u) \subset k - \widetilde{U}ST_s(\lambda, q, \nu, \varsigma)$.

Proof. It is clear that $u \in k - \widetilde{U}ST_s(\lambda, q, \nu, \varsigma)$ if and only if

$$\left| \frac{(1-\varsigma)z\left(\mathscr{R}_q^\lambda u(z)\right)'(1+ke^{i\varrho})-(ke^{i\varrho}+1+\nu)\left(\mathscr{R}_q^\lambda u(z)-\mathscr{R}_q^\lambda u(\varsigma z)\right)}{(1-\varsigma)z\left(\mathscr{R}_q^\lambda u(z)\right)'(1+ke^{i\varrho})+(1-ke^{i\varrho}-\nu)\left(\mathscr{R}_q^\lambda u(z)-\mathscr{R}_q^\lambda u(\varsigma z)\right)} \right| < 1, \qquad -\pi < \varrho < \pi.$$

For any complex number s ($|s| = 1$) we may write

$$\frac{(1-\varsigma)z\left(\mathscr{R}_q^\lambda u(z)\right)'(1+ke^{i\varrho})-(ke^{i\varrho}+1+\nu)\left(\mathscr{R}_q^\lambda u(z)-\mathscr{R}_q^\lambda u(\varsigma z)\right)}{(1-\varsigma)z\left(\mathscr{R}_q^\lambda u(z)\right)'(1+ke^{i\varrho})+(1-ke^{i\varrho}-\nu)\left(\mathscr{R}_q^\lambda u(z)-\mathscr{R}_q^\lambda u(\varsigma z)\right)} \neq s.$$

That is,

$$(1-s)(1-\varsigma)z\left(\mathscr{R}_q^\lambda u(z)\right)'(1+ke^{i\varrho}) - (ke^{i\varrho}+1+\nu+s(-1+ke^{i\varrho}+\nu) \times \left(\mathscr{R}_q^\lambda u(z)-\mathscr{R}_q^\lambda u(\varsigma z)\right) \neq 0$$

$$\Rightarrow z - \sum_{\eta=2}^{\infty} \frac{\phi_\eta(\lambda, q)[(\eta - u_\eta)(1+ke^{i\varrho} - ske^{i\varrho}) - s(\eta + u_\eta) - u_\eta\nu(1-s)]}{\nu(s-1) - 2s} z^\eta \neq 0.$$

However, $u \in k - \widetilde{U}ST_s(\lambda, q, \nu, \varsigma) \Leftrightarrow \frac{(u*h)}{z} \neq 0 \ (z \in \Delta - \{0\})$, where

$$h(z) = z - \sum_{\eta=2}^{\infty} c_\eta z^\eta$$

and

$$c_\eta = \frac{\phi_\eta(\lambda, q)\left[(\eta - u_\eta)(1+ke^{i\varrho} - ske^{i\varrho}) - s(\eta + u_\eta) - u_\eta\nu(1-s)\right]}{\nu(s-1) - 2s}.$$

Since $\frac{u(z)+\epsilon z}{1+\epsilon} \in k - \widetilde{U}ST_s(\lambda, q, \nu, \varsigma)$, we observe that

$$|c_\eta| \leq \frac{\phi_\eta(\lambda, q)\,|\eta(1+k) - u_\eta(k+\nu)|}{1-\nu}.$$

Therefore, $z^{-1}\left(\frac{u(z)+\epsilon z}{1+\epsilon} * h(z)\right) \neq 0$, which is equivalent to

$$\frac{(u*h)(z)}{(1+\epsilon)z} + \frac{\epsilon}{1+\epsilon} \neq 0. \tag{7.13}$$

Now, let us consider that $\left|\frac{(u*h)(z)}{z}\right| < \upsilon$. From Equation (7.13), we get

$$\begin{aligned}\left|\frac{(u*h)(z)}{(1+\epsilon)z} + \frac{\epsilon}{1+\epsilon}\right| &\geq \frac{|\epsilon|}{|1+\epsilon|} - \frac{1}{|1+\epsilon|}\left|\frac{(u*h)(z)}{z}\right| \\ &> \frac{|\epsilon|-\upsilon}{|1+\epsilon|} \geq 0.\end{aligned}$$

This contradicts that $|\epsilon| < \upsilon$, and, hence, we have $\left|\frac{(u*h)(z)}{z}\right| \geq \upsilon$.

If $g(z) = z - \sum\limits_{\eta=2}^{\infty} b_\eta z^\eta \in N_\upsilon(u)$, then

$$\begin{aligned}\upsilon - \left|\frac{(g*h)(z)}{z}\right| &\leq \left|\frac{((u-g)*h)(z)}{z}\right| \\ &\leq \sum_{\eta=2}^{\infty} |a_\eta - b_\eta|\,|c_\eta|\,|z|^\eta \\ &< \sum_{\eta=2}^{\infty} \frac{\phi_\eta(\lambda,q)\,|\eta(1+k) - u_\eta(k+\nu)|}{1-\nu}\,|a_\eta - b_\eta| \\ &\leq \upsilon.\end{aligned}$$

□

7.4 Partial Sums

In this section, employing a method exploited by Silverman [13], we survey the proportion of a mapping $u \in T$ to its sequence of partial sums $u_m(z) = z + \sum\limits_{\eta=2}^{m} a_\eta z^\eta$.

Theorem 7.3. *If the function $u \in T$ fulfills (7.11) then*

$$\Re\left\{\frac{u(z)}{u_m(z)}\right\} \geq 1 - \frac{1}{\chi_{m+1}}$$

and

$$\chi_\eta = \begin{cases} 1, & \eta = 2,\ldots,m \\ \chi_{m+1}, & \eta = m+1, m+2,\ldots \end{cases},$$

where

$$\chi_\eta = \frac{\phi_\eta(\lambda, q)|\eta(1+k) - u_\eta(k+\nu)|}{1-\nu}.$$

For every m, the estimate is sharp with

$$u(z) = z + \frac{z^{m+1}}{\chi_{m+1}}. \tag{7.14}$$

Proof. Suppose that

$$\begin{aligned} \tfrac{1+w(z)}{1-w(z)} &= \chi_{m+1}\left\{\tfrac{u(z)}{u_m(z)} - \left(1 - \tfrac{1}{\chi_{m+1}}\right)\right\} \\ &= \left\{\frac{1+\sum\limits_{\eta=2}^{m} a_\eta z^{\eta-1} + \chi_{m+1}\sum\limits_{\eta=m+1}^{\infty} a_\eta z^{\eta-1}}{1+\sum\limits_{\eta=2}^{m} a_\eta z^{\eta-1}}\right\}. \end{aligned} \tag{7.15}$$

Then, from Equation (7.15), we have

$$w(z) = \frac{\chi_{m+1}\sum\limits_{\eta=m+1}^{\infty} a_\eta z^{\eta-1}}{2\left(1+\sum\limits_{\eta=2}^{m} a_\eta z^{\eta-1}\right) + \chi_{m+1}\sum\limits_{\eta=m+1}^{\infty} a_\eta z^{\eta-1}}$$

and

$$|w(z)| \leq \frac{\chi_{m+1}\sum\limits_{\eta=m+1}^{\infty} a_\eta}{2 - 2\sum\limits_{\eta=2}^{m} a_\eta - \chi_{m+1}\sum\limits_{\eta=m+1}^{\infty} a_\eta}.$$

Next, $|w(z)| \leq 1$ if

$$\begin{aligned} 2\chi_{m+1}\sum_{\eta=m+1}^{\infty} a_\eta &\leq 2\left(1 - \sum_{\eta=2}^{m} a_\eta\right) \\ \Rightarrow \sum_{\eta=2}^{m} a_\eta + \chi_{m+1}\sum_{\eta=m+1}^{\infty} a_\eta &\leq 1. \end{aligned} \tag{7.16}$$

It is enough to show that the left-hand side of Equation (7.16) is bounded above by $\sum_{\eta=2}^{\infty} \chi_\eta a_\eta$, that is,

$$\sum_{\eta=2}^{m}(\chi_\eta - 1)a_\eta + \sum_{\eta=m+1}^{\infty}(\chi_\eta - \chi_{m+1})a_\eta \geq 0.$$

To find the sharp result, for $z = re^{i\pi/\eta}$, we have

$$\frac{u(z)}{u_m(z)} = 1 + \frac{z^m}{\chi_{m+1}}.$$

Thus, by taking $z \to 1^-$, we get

$$\frac{u(z)}{u_m(z)} = 1 - \frac{1}{\chi_{m+1}}.$$

□

We now express bounds for $\frac{u_m(z)}{u(z)}$.

Theorem 7.4. *If the function* $u \in T$ *fulfills Equation* (7.11), *then*

$$\Re\left\{\frac{u_m(z)}{u(z)}\right\} \geq \frac{\chi_{m+1}}{1 + \chi_{m+1}}.$$

The estimate is sharp for Equation (7.14).

Proof. It is a routine to verify that

$$\frac{1+w(z)}{1-w(z)} = (1+\chi_{m+1})\left\{\frac{u_m(z)}{u(z)} - \frac{\chi_{m+1}}{1+\chi_{m+1}}\right\}$$

$$\left\{\frac{1+\sum_{\eta=2}^{m} a_\eta z^{\eta-1} - \chi_{m+1}\sum_{\eta=m+1}^{\infty} a_\eta z^{\eta-1}}{1+\sum_{\eta=2}^{\infty} a_\eta z^{\eta-1}}\right\},$$

where

$$w(z) = \frac{(1+\chi_{m+1})\sum_{\eta=m+1}^{\infty} a_\eta z^{\eta-1}}{-2\left(1+\sum_{\eta=2}^{m} a_\eta z^{\eta-1}\right) - (1-\chi_{m+1})\sum_{\eta=m+1}^{\infty} a_\eta z^{\eta-1}}.$$

It follows that

$$|w(z)| \leq \frac{(1+\chi_{m+1}) \sum\limits_{\eta=m+1}^{\infty} a_\eta}{2 - 2\sum\limits_{\eta=2}^{m} a_\eta + (1-\chi_{m+1}) \sum\limits_{\eta=m+1}^{\infty} a_\eta} \leq 1$$

$$\Rightarrow \sum_{\eta=2}^{m} a_\eta + \chi_{m+1} \sum_{\eta=m+1}^{\infty} a_\eta \leq 1. \tag{7.17}$$

It is enough to express that LHS of Equation (7.17) is bounded above by $\sum\limits_{\eta=2}^{\infty} \chi_\eta a_\eta$, which is equivalent to

$$\sum_{\eta=2}^{m} (\chi_\eta - 1) a_\eta + \sum_{\eta=m+1}^{\infty} (\chi_\eta - \chi_{m+1}) a_\eta \geq 0.$$

□

Theorem 7.5. *If the function u of the form* (7.1) *fulfills* (7.11), *then*

$$\Re\left\{\frac{u'(z)}{u'_m(z)}\right\} \geq 1 - \frac{m+1}{\chi_{m+1}} \tag{7.18}$$

and

$$\Re\left\{\frac{u'_m(z)}{u'(z)}\right\} \geq \frac{\chi_{m+1}}{1+m+\chi_{m+1}},$$

where

$$\chi_\eta \geq \begin{cases} 1, & \eta = 1, 2, \ldots, m \\ \eta \frac{\chi_{m+1}}{m+1}, & \eta = m+1, m+2, \ldots \end{cases}.$$

These estimates are sharp (7.14).

Proof. For u given by Equation (7.1), we may write

$$\begin{aligned} \tfrac{1+w(z)}{1-w(z)} &= \chi_{m+1}\left\{\frac{u'(z)}{u'_m(z)} - \left(1 - \frac{m+1}{\chi_{m+1}}\right)\right\} \\ &= \left\{1 + \sum_{\eta=2}^{m} \eta a_\eta z^{\eta-1} + \tfrac{\chi_{m+1}}{m+1} \sum_{\eta=m+1}^{\infty} \eta a_\eta z^{\eta-1} 1 + \sum_{\eta=2}^{m} a_\eta z^{\eta-1}\right\}, \end{aligned}$$

where

$$w(z) = \frac{\frac{\chi_{m+1}}{m+1} \sum\limits_{\eta=m+1}^{\infty} \eta a_\eta z^{\eta-1}}{2 + 2\sum\limits_{\eta=2}^{m} \eta a_\eta z^{\eta-1} + \frac{\chi_{m+1}}{m+1} \sum\limits_{\eta=m+1}^{\infty} \eta a_\eta z^{\eta-1}}.$$

Then, we have

$$|w(z)| \le \frac{\frac{\chi_{m+1}}{m+1} \sum\limits_{\eta=m+1}^{\infty} \eta a_\eta}{2 - 2\sum\limits_{\eta=2}^{m} \eta a_\eta + \frac{\chi_{m+1}}{m+1} \sum\limits_{\eta=m+1}^{\infty} \eta a_\eta}.$$

From the above inequality, we get

$$|w(z)| \le 1 \Leftrightarrow \sum_{\eta=2}^{m} \eta a_\eta + \frac{\chi_{m+1}}{m+1} \sum_{\eta=m+1}^{\infty} \eta a_\eta \le 1, \tag{7.19}$$

since the LHS of Equation (7.19) is bounded above by $\sum\limits_{\eta=2}^{\infty} \chi_\eta a_\eta$.

By using the same method as before, we also obtain Equation (7.18). □

References

[1] E. Aqlan, J. M. Jhangiri and S.R. Kulkarni, *Classes of $k-$uniformly convex and starlike functions,* Tamkang J. Math. **35**, 261–266, 2004.

[2] A.W. Goodman, *Univalent functions and nonanalytic curves*, Proc. Amer. Math. Soc. **8**, 598–601, 1957.

[3] A.W. Goodman, *On uniformly starlike functions*, J. Math. Anal. Appl. **155**, 364–370, 1991.

[4] M. L. Morga, *Application of Ruscheweyh derivatives and Hadamard product to analytic function,* Int. J. Math. Math. Sci. **22** (4), 795–805, 1999.

[5] K. L. Noor and S. Hussain, *On certain analytic function associated with Ruscheweyh derivatives and bounded Mocanu variation,* J. Math. Anal. **34** (2), 1145–1152, 2008.

[6] S. Owa, T. Sekine and R. Yamakawa, *On Sakaguchi type functions*, Appl. Math. Comput. **187**, 356–361, 2007.

[7] S. Ruscheweyh, *Neighbourhoods of univalent functions*, Proc. Amer. Math. Soc. **81** (4), 521–527, 1981.

[8] S. Ruscheweyh, *New criteria for univalent functions,* Proc. Amer. Math. Soc. **49** , 109–115, 1975.

[9] M. P. Santosh, R.N. Ingle, P. Thirupathi Reddy and B. Venkateswarlu, *A new subclass of analytic functions defined by linear operator*, Adv. in Math. Sci. J. **9**(1), 205–217, 2020.

[10] K. Sakaguchi, *On a certain univalent mapping*, J. Math. Soc. Japan. **11**, 72–75, 1959.

[11] S. L. Shukla and V. Kumar, *Univalent functions defined by Ruscheweyh derivatives,* Int. J. Math. Math. Sci. **6**, 483–486, 1983.

[12] H. Silverman, *Univalent functions with negative coefficients*, Proc. Amer. Math. Soc. **51**, 109–116, 1975.

[13] H. Silverman, *Partial sums of starlike and convex functions*, J. Anal. Appl. **209**, 221–227, 1997.

8

Second Law Analysis of Williamson Nanofluid Flow in Attendance of Radiation and Heat Generation with Cattaneo–Christov (C–C) Heat Flux

K. Loganathan[1], C. Selvamani[2], and A. Charles Sagayaraj[3]

[1]Research and Development Wing, Live4Research, Tiruppur, Tamil Nadu 638106, India
[2]Department of Mathematics, Faculty of Engineering, Karpagam Academy of Higher Education, Coimbatore, Tamil Nadu 641021, India
[3]Department of Mathematics, Sri Vidya Mandir Arts and Science College, Katteri, Uthangarai, Tamil Nadu, India
E-mail: loganathankaruppusamy304@gmail.com

Abstract

This chapter investigates the thermal radiation and convective surface, heating effects incorporated into magneto hydrodynamic Williamson nanofluid flow concerning stretchable sheet. Classical Fourier heat flux was replaced by Cattaneo–Christov (C–C) heat flux. The entropy rate of Williamson nanomaterial is also computed. The physical governing systems are transformed into nonlinear ordinary system by exploiting similarity variables. The nonlinear ordinary system is verified through homotopy scheme. Upshots of the different flow parameters, Bejan number, and entropy generation of distinct parameters were examined. Moreover, mass and heat transmission rates are computed and studied.

Keywords: Cattaneo–Christov (C–C) heat flux, Brownian motion, Williamson nanofluid, entropy generation, homotopy analytic scheme.

8.1 Introduction

The study of non-Newtonian fluid flows with heat and mass transfer were given more consideration in past years. Molla and Yao [1] explored the heat transmission of non-Newtonian fluids toward a stretchy surface using an altered power-law model. They presented numerical computations of shear-thinning fluid with velocity and temperature fields. Loganathan *et al.* [2] presented the cross-diffusion and radiation effects of Oldroyd-B type fluid bounded by a stretchable surface with second-order slip effects. They solve the dimensionless equations through a homotopy analytic procedure. Sahoo [3] considered the slip flow of third-grade fluid caused by a linear stretchy sheet with heat transmission effects. He implemented a numerical system to resolve the differential equations and find the impacts of third-grade liquid parameters on velocity and temperature fields. Goyal and Bhargava [4] studied the impressions of heat transfer of viscoelastic nanofluid induced by a stretchy plate due to heat source/sink. Khan *et al.* [5] examined the viscoelastic model unsteady stagnation point flow bounded by a stretchable plate. They solved the nonlinear equations with the help of homotopy analysis method. Other notable research see through the non-Newtonian fluid flows in [6–10].

In previous years, several scientists had investigated the entropy generation in fluid flow and heat transmission toward a stretchable surface. Ijaz *et al.* [11] analyzed the entropy computation of Williamson nano material flow over a shrinking surface with joule heating and chemical reaction effects. They found the solution of nonlinear equations through optimal Homotopy analysis method and calculated the impacts of entropy generation and Bejan number. Baag *et al.* [12] explored the magneto-convective flow of Walter-B liquid embedded in a porous plate with irreversibility analysis. Flow ranges are tackled by applying the Kummer function. Entropy generation has been found to implement the Joule heating and dissipation. Nemat [13] presented the entropy analysis of forced convection flow of Jeffrey liquid bounded by a stretchy plate. Moreover, entropy generation enhances with the improvement of Deborah number. Abolbashari *et al.* [14] had examined the entropy generation and heat and mass transfer for laminar flow for nanofluid bounded by stretchy surface in the appearance of slip and convective conditions. Butt *et al.* [15] found the entropy properties of MHD flow and heat transmission toward a stretchy cylinder. The prominent related surveys in this research area are presented in [16–19].

Inspired by the above works of literature, we construct the two-dimensional steady flow of Williamson nanofluid with radiation and convective heating conditions. Cattaneo–Christov (C–C) heat flux is added instead of ordinary heat flux. Entropy generation analysis is made for different fluid parameters. Also, Bejan number is calculated for various physical parameters. Nonlinear physical systems are computed via HAM technique [20–25].

8.2 Problem Development

The examination of this current problem is focused on entropy optimization of Williamson nano material flow through stretchy sheet with heated surface conditions and radiation. Modified heat flux is incorporated. The governing physical systems along with boundary conditions are shown in Figure 8.1.

$$\frac{\partial U}{\partial X}+\frac{\partial V}{\partial Y}=0 \tag{8.1}$$

$$U\frac{\partial U}{\partial X}+V\frac{\partial U}{\partial Y}-\sqrt{2}\,\nu\beta\,\frac{\partial U}{\partial Y}\frac{\partial^2 U}{\partial Y^2}-\frac{\sigma B_0^2}{\rho}U=\nu\frac{\partial^2 U}{\partial Y^2} \tag{8.2}$$

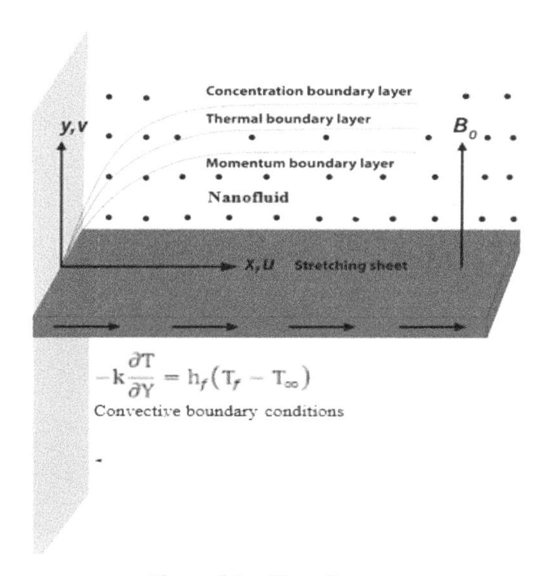

Figure 8.1 Flow diagram.

$$\mathrm{U}\frac{\partial \mathrm{T}}{\partial \mathrm{X}}+\mathrm{V}\frac{\partial \mathrm{T}}{\partial \mathrm{Y}}+\frac{1}{\rho \mathrm{c_p}}\frac{\partial \mathrm{q_r}}{\partial \mathrm{Y}}-\frac{\mathrm{Q_0}}{\rho \mathrm{c_p}}(\mathrm{T}-\mathrm{T_\infty})=\frac{\mathrm{k}}{\rho \mathrm{c_p}}\frac{\partial^2 \mathrm{T}}{\partial \mathrm{Y}^2}$$
$$+\tau\left[\mathrm{D_B}\frac{\partial \mathrm{C}}{\partial \mathrm{Y}}\frac{\partial \mathrm{T}}{\partial \mathrm{Y}}+\frac{\mathrm{D_T}}{\mathrm{T_\infty}}\left(\frac{\partial \mathrm{T}}{\partial \mathrm{Y}}\right)^2\right]+\lambda_T\left(U^2\frac{\partial^2 T}{\partial X^2}+V^2\frac{\partial^2 T}{\partial Y^2}+\right.$$
$$\left.\left(U\frac{\partial U}{\partial X}\frac{\partial T}{\partial X}+V\frac{\partial U}{\partial Y}\frac{\partial T}{\partial X}\right)+2UV\frac{\partial T^2}{\partial X\partial Y}\right) \tag{8.3}$$

$$\mathrm{U}\frac{\partial \mathrm{C}}{\partial \mathrm{X}}+\mathrm{v}\frac{\partial \mathrm{C}}{\partial \mathrm{Y}}=\mathrm{D_B}\frac{\partial^2 \mathrm{C}}{\partial \mathrm{Y}^2}+\frac{\mathrm{D_T}}{\mathrm{T_\infty}}\frac{\partial^2 \mathrm{T}}{\partial \mathrm{Y}^2} \tag{8.4}$$

$$\mathrm{U}=\mathrm{U_w}(\mathrm{X})=\mathrm{aX},\ -\mathrm{k}\frac{\partial \mathrm{T}}{\partial \mathrm{Y}}=\mathrm{h}_f(\mathrm{T}_f-\mathrm{T_\infty}),\ \mathrm{T}=\mathrm{T_w},\ \mathrm{C}=\mathrm{C_w}\ \text{at}\ \mathrm{Y}=0$$
$$\mathrm{U}\to 0,\ \mathrm{T}\to\mathrm{T_\infty},\ \mathrm{C}\to\mathrm{C_\infty}\ \text{as}\ \mathrm{Y}\to\infty. \tag{8.5}$$

The similarity transformations are

$$\psi=\sqrt{a\nu}Xf(\eta),\ U=\frac{\partial\psi}{\partial Y},\ V=-\frac{\partial\psi}{\partial X},\ \eta=\sqrt{\frac{a}{\nu}}Y,\ \theta(\eta)=\frac{T-T_\infty}{T_f-T_\infty},$$
$$V=-\sqrt{a\nu}f(\eta),\ U=aXf'(\eta),\ \phi(\eta)=\frac{C-C_\infty}{C_w-C_\infty}. \tag{8.6}$$

We have,

$$f'''+Wef''f'''-f'^2+ff''-Mf'=0 \tag{8.7}$$

$$\left(1+\frac{4}{3}Rd\right)\theta''+Prf\theta'+PrS\theta-Pr\gamma\left(ff'\theta'+f^2\theta''\right)+PrNb\,\theta'\phi'$$
$$+PrNt\theta'^2=0 \tag{8.8}$$

$$\phi^\phi+Lef\phi'+\frac{Nt}{Nb}\theta''=0 \tag{8.9}$$

with the end points

$$f(\eta)=f_w,\ f'(\eta)=1,\ \phi(\eta)=1,\ \theta'(\eta)=-Bi(1-\theta(0))\ \text{at}\ \eta=0$$
$$f'(\eta)\to 0,\ \phi(\eta)\to 0,\ \theta(\eta)\to 0,\ \text{as}\ \eta\to\infty. \tag{8.10}$$

The non-dimensional variables are declared as Weissenberg number $(We) = \sqrt{\frac{2a^3}{\upsilon}}\beta\, X$, magnetic field parameter $M = \sigma B_0^2/\rho a$, Prandtl number $(Pr) = \rho C_p/k$, radiation parameter $(Rd) = \left(4\sigma^* T_\infty^3\right)/(kk^*)$, heat generation parameter $(S) = \frac{Q_0}{\rho c_p}$, thermal relaxation time $(\gamma) = \lambda_T a$, Biot number $(Bi) = \frac{h_f}{k}\sqrt{\nu/a}$, Brownian motion parameter $(Nb) = \frac{\tau D_B}{\nu}(C_w - C_\infty)$, and thermophoresis parameter $(Nt) = \frac{\tau D_T}{\nu}(T_w - T_\infty)$.

The physical entities are defined as

$$C_f Re^{0.5} = 2\left[f''(0) + f''^2 \frac{We}{2}(0)\right]$$

$$NuRe^{-0.5} = -\left(1 + \frac{4}{3}Rd\right)\theta'(0)$$

$$ShRe^{-0.5} = -\phi'(0).$$

8.3 Formulation of Entropy Generation

The entropy of the physical structure is stated as

$$S_{gen}''' = \frac{K_1}{T_\infty^2}\left[\left(\frac{\partial T}{\partial X}\right)^2 + \left(\frac{\partial T}{\partial Y}\right)^2 + \frac{16\sigma^* T_\infty^3}{3kk^*}\left(\frac{\partial T}{\partial Y}\right)^2\right] +$$
$$\frac{\mu}{T_\infty}\left[\left(\frac{\partial U}{\partial Y}\right)^2 + \beta\left(\frac{\partial U}{\partial Y}\right)^3\right] + \left[\frac{\partial U}{\partial Y} + \frac{\partial V}{\partial X}\right]^2$$
$$+ \frac{RD}{C_\infty}\left[\left(\frac{\partial C}{\partial Y}\right)^2\right] + \frac{RD}{T_\infty}\left[\left(\frac{\partial T}{\partial Y}\right)\left(\frac{\partial C}{\partial Y}\right)\right] + \frac{\sigma B_0^2}{T_\infty}U^2. \tag{8.11}$$

The dimensionless form becomes

$$E_G = Re\left(1 + \frac{4}{3}Rd\right)\theta'^2 + Re\frac{Br}{\Omega}\left(f''^2 + \frac{We}{\sqrt{2}}f''^3\right) + \frac{Br}{\Omega}Mf'^2 +$$
$$Re\left(\frac{\zeta}{\Omega}\right)^2\lambda\phi'^2 + Re\frac{\zeta}{\Omega}\lambda\phi'\theta'. \tag{8.12}$$

The Bejan number states as

$$\text{Be} = \frac{Re\left(1 + \frac{4}{3}Rd\right)\theta'^2 + Re\left(\frac{\zeta}{\Omega}\right)^2\lambda\phi'^2 + Re\frac{\zeta}{\Omega}\lambda\phi'\theta'}{\begin{array}{c}Re\left(1 + \frac{4}{3}Rd\right)\theta'^2 + Re\frac{Br}{\Omega}\left(f''^2 + \frac{We}{\sqrt{2}}f''^3\right) + \\ Re\left(\frac{\zeta}{\Omega}\right)^2\lambda\phi'^2 + Re\frac{\zeta}{\Omega}\lambda\phi'\theta' + \frac{Br}{\Omega}Mf'^2\end{array}}. \tag{8.13}$$

8.4 HAM Solutions

The primary assumptions for homotopy technique are stated as $f_0 = f_w + 1 - e^{-\eta}$, $\theta_0 = \frac{Bi * e^{-\eta}}{1+Bi}$ and $\phi_0 = e^{-\eta}$. The auxiliary linear operators $\mathcal{L}_f$, $\mathcal{L}_\theta$, and $\mathcal{L}_\phi$ are derived as $\mathcal{L}_f = [f^{'''}(\eta) - f^{'}(\eta)]$, $\mathcal{L}_\theta = [\theta^{''}\eta - \theta(\eta)]$, and $\mathcal{L}_\phi = [\phi^{''}(\eta) - \phi(\eta)]$ with adequate following properties $\mathcal{L}_f[A_1 + A_2 e^{\eta} + e^{-\eta} A_3] = 0$, $\mathcal{L}_\theta[A_4 e^{\eta} + A_5 e^{-\eta}] = 0$, and $\mathcal{L}_\phi[A_6 e^{\eta} + A_7 e^{-\eta}]$ where $A_j\,(j = 1 - 7)$ denotes the arbitrary conditions.

The special solutions $[f_m^*, \theta_m^*, \phi_m^*]$ are

$$f_m(\eta) = f_m^*(\eta) + A_1 + A_2 e^{\eta} + A_3 e^{-\eta}$$
$$\theta_m(\eta) = \theta_m^*(\eta) + A_4 e^{\eta} + A_5 e^{-\eta}$$
$$\phi_m(\eta) = \phi_m^*(\eta) + A_6 e^{\eta} + A_7 e^{-\eta}.$$

The auxiliary constants h_f, h_θ, and h_ϕ hold a vital role in convergence solutions. The h-charts of $f^{''}(0)$, $\theta^{'}(0)$, and $\phi^{'}(0)$ for We, γ, and Nb are shown in Figure 8.2. From these curves, straight line is referred as the h-curve. Table 8.1 indicates the series solution is attained for 15th order of approximations.

8.5 Computational Results and Discussion

Upshots of entropy rate, Bejan number, nanoparticle concentration, temperature, and velocity distributions are expressed in this section. The impacts of Weissenberg number (We) on $f^{'}(\eta)$ are sketched in Figure 8.3. It is noted that enhancing the value of relaxation time of liquid tends to reduce the velocity of the fluid flow. The higher magnetic field parameter M decays the velocity profile $f^{'}(\eta)$ (see Figure 8.4).

The influence of temperature profile $\theta(\eta)$ associated with numerous ranges of radiation constant is examined in Figure 8.5. In Figure 8.5, we note that higher temperatures and thicker thermal boundary layers are found from larger thermal radiation (Rd) constant. An increase in temperature is noted that, due to the higher radiation, produces an extra heat to the liquid. The radiation effects are more pronounced for ordinary heat flux compared to the C-C heat flux. Figure 8.6 displayed Nt influence on concentration. Clearly, concentration enhances for higher Nt. This is due to the fast flow through the stretching sheet transfer with its nano particles taking the lead to an increase in boundary layer of mass volume fraction

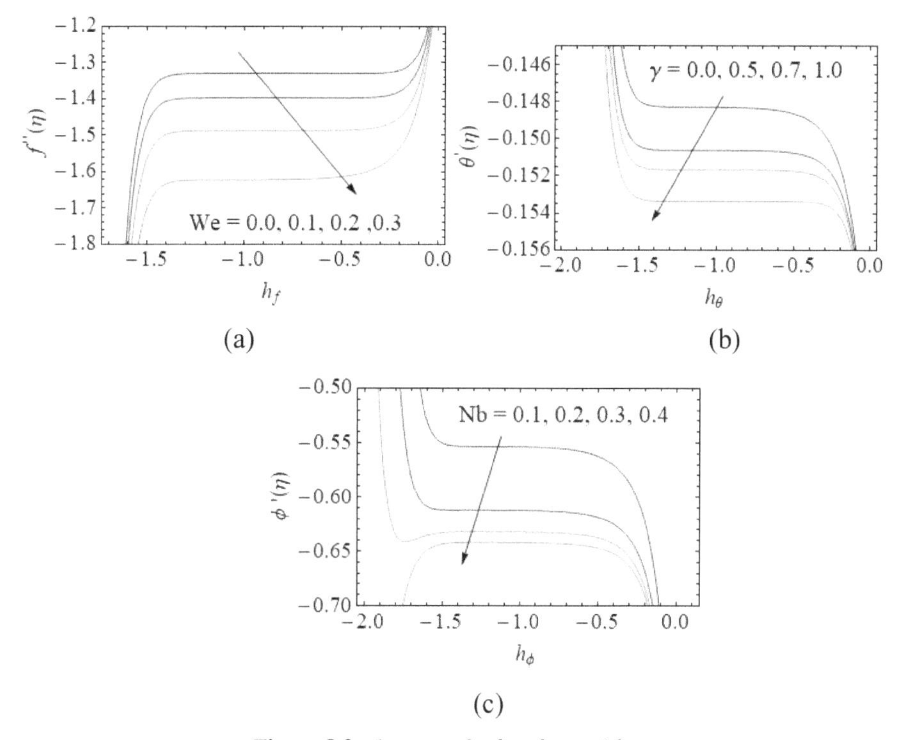

Figure 8.2 h-curves for h_f, h_θ, and h_ϕ.

Table 8.1 Order of approximations of HAM.

Order	$-f''(0)$	$-\theta'(0)$	$-\phi'(0)$	CPU Time (s)
2	1.3491	0.1551	0.6869	0.61
7	1.3959	0.1510	0.5761	5.672
12	1.3966	0.1506	0.5588	25.046
16	1.3966	0.1506	0.5546	63.609
22	1.3966	0.1506	0.5535	216.73
27	1.3966	0.1506	0.5531	617.73
31	1.3966	0.1506	0.5531	1229.70
35	1.3966	0.1506	0.5531	2353.78

Figures 8.7–8.10 highlight the entropy generation effects of various parameters. Figure 8.7 displays the entropy generation rate with Weissenberg number. Figures 8.8 and 8.9 show characteristics of magnetic parameter (M) and thermal relaxation time (γ) on E_G. Here, entropy increases for higher assessment of M and γ. Figure 8.8 detected that an extra traverse magnetic

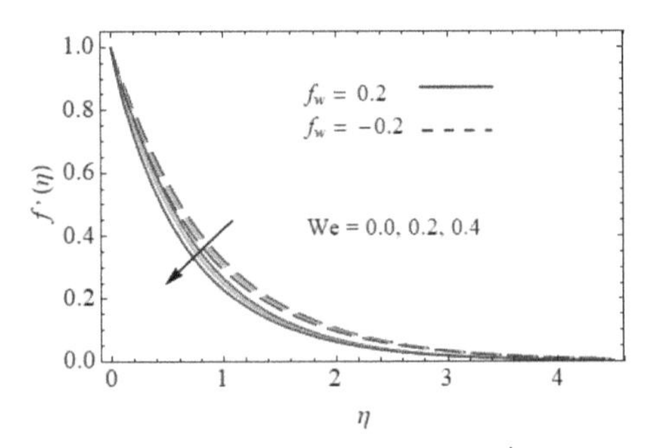

Figure 8.3 Impact of We on $f'(\eta)$.

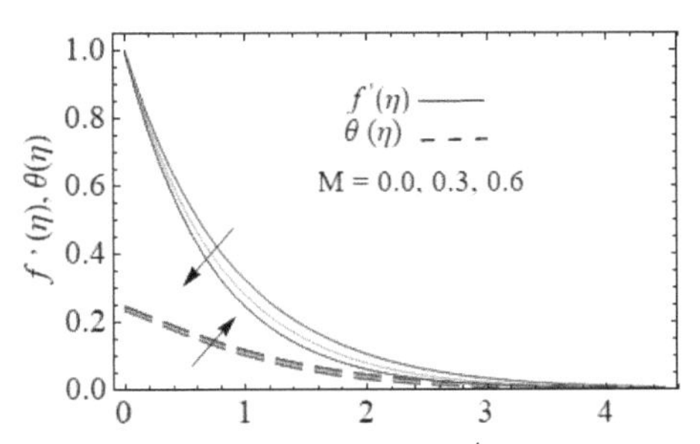

Figure 8.4 Impact of M on $f'(\eta)$ and $\theta(\eta)$.

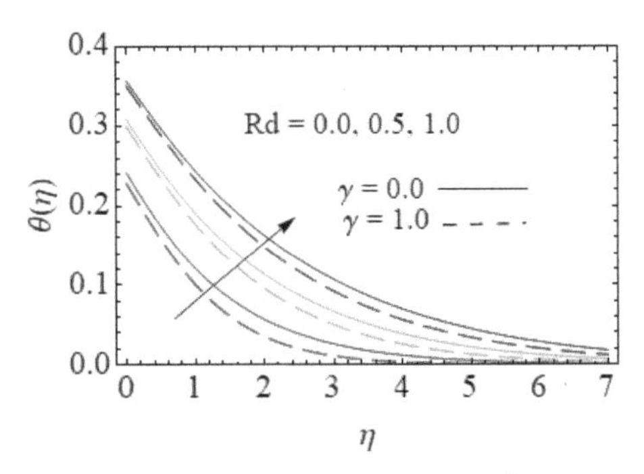

Figure 8.5 Impact of Rd on $\theta(\eta)$.

field tends to diminish the E_G. From Figure 8.9, we know that thermal relaxation time is low for temperature and heat transfer rates. Moreover, the domination of the irreversibility in heat transfer affected the heat flux. Thus, we have seen a small increase in the entropy of the system (see Figures 8.10 and 8.11). It is noted that the larger range of Br delivers the opposite impacts in entropy and Bejan number profiles.

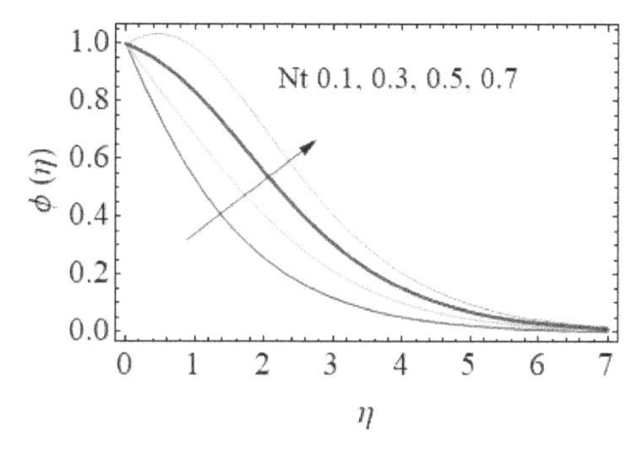

Figure 8.6 Impact of Nt on $\phi(\eta)$.

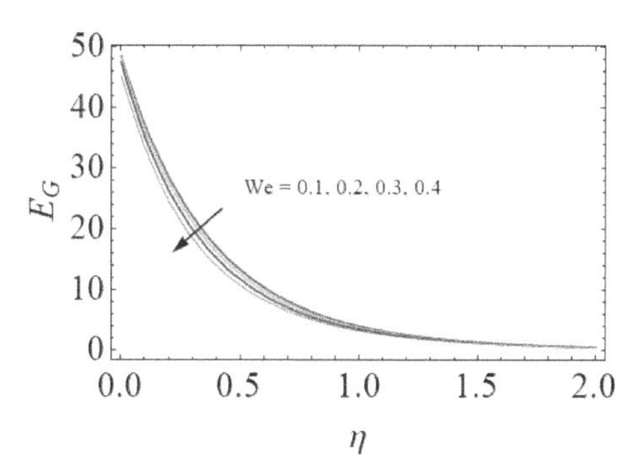

Figure 8.7 Impact of We on E_G.

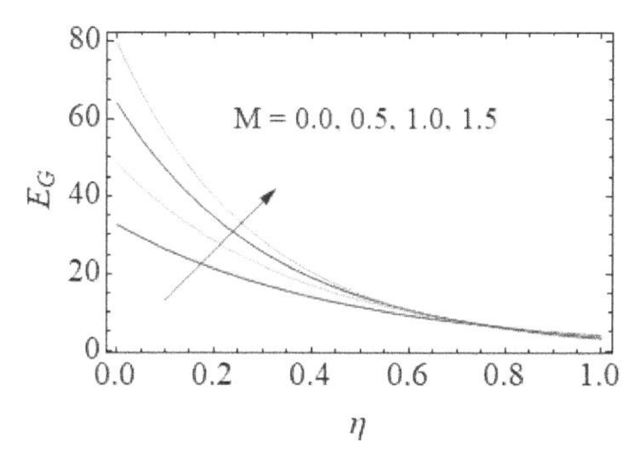

Figure 8.8 Impact of on E_G.

From Figure 8.12, the Be increases, where as the growing values of (We). The significance of thermal relaxation time (γ) on Be is discussed in Figure 8.13. From this graph, we note that in the beginning, Be expands for $(\gamma = 0.0, 0.3, 0.6, 0.9)$. After $\eta > 2.2$, consequently, Be reduces.

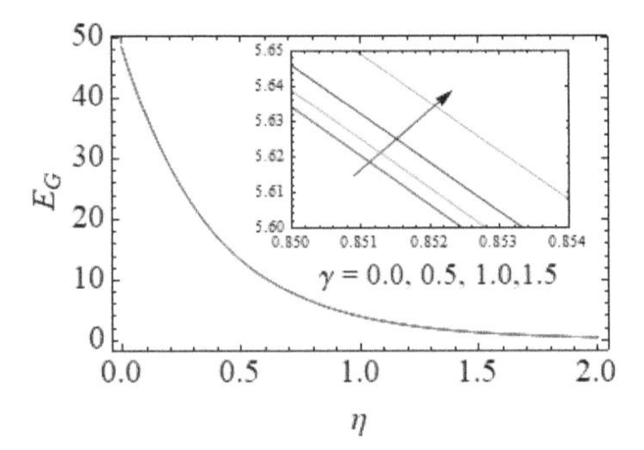

Figure 8.9 Impact of γ on E_G.

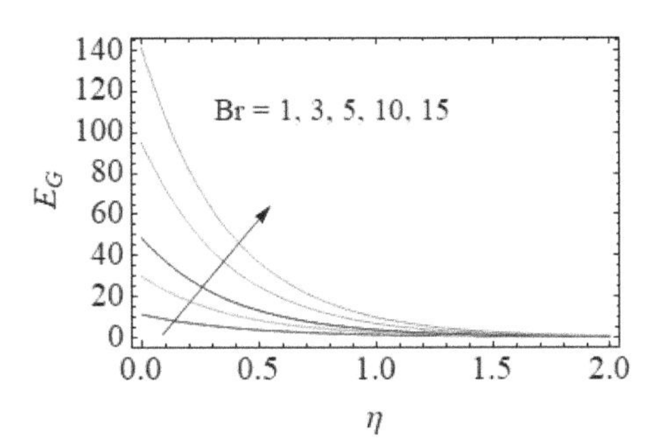

Figure 8.10 Impact of Br on E_G.

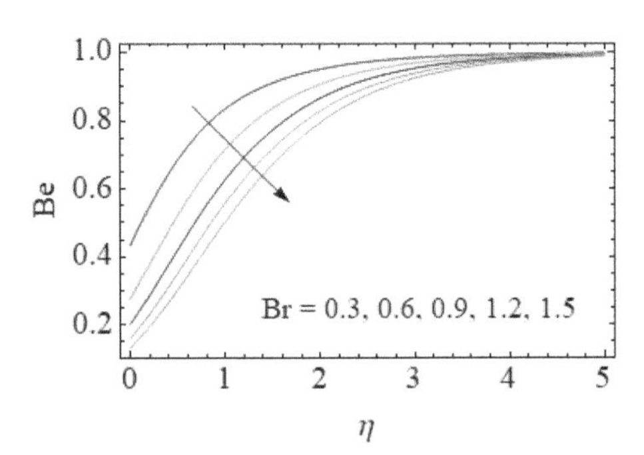

Figure 8.11 Impact of Br on Be.

Figure 8.14 discusses the impacts of the magnetic field (M) on the Be. From Figure 8.14, we note that Be falls initially and then begins to rise subsequently reaching the value of $\eta > 1$.

Influence of diverse fluid parameters on surface drag force, local heat, and mass transmission ratios is demonstrated in Figures 8.15–8.18.

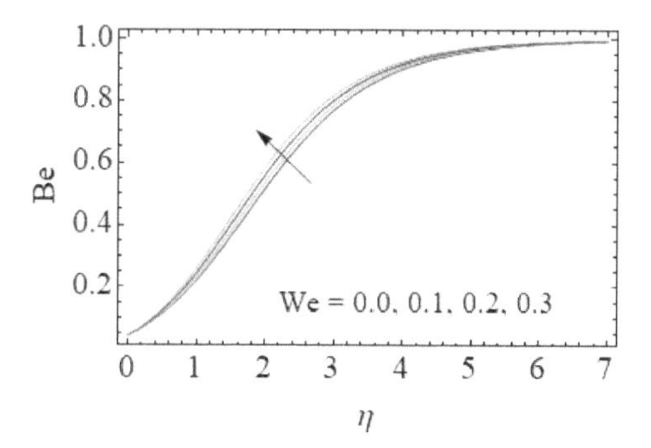

Figure 8.12 Impact of We on Be.

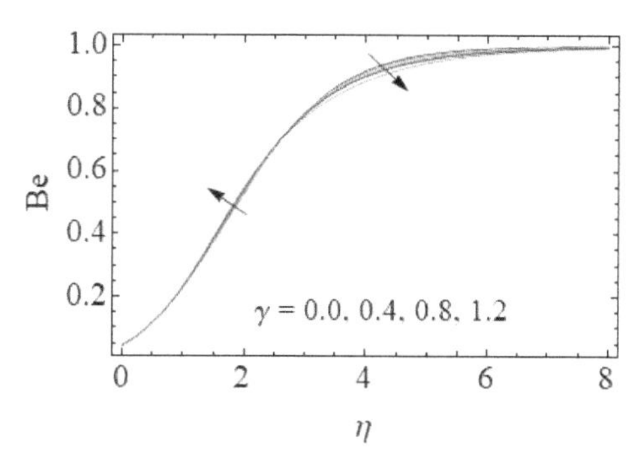

Figure 8.13 Impact of γ on Be.

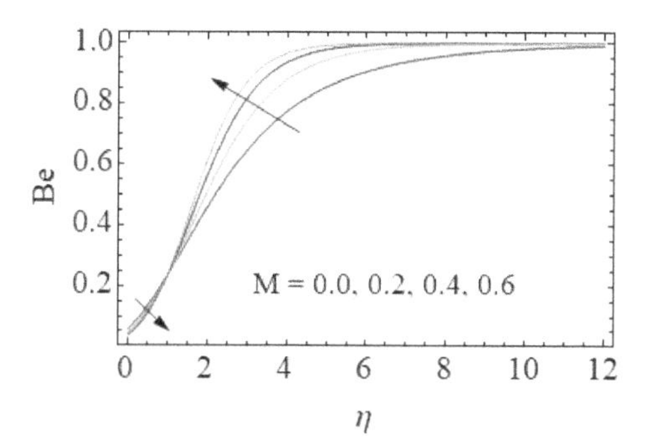

Figure 8.14 Impact of M on Be.

From Figures 8.15 and 8.16, we obtained that surface drag force diminishes for the upsurge values of the merged parameters We and f_w and We and M. Figures 8.17 and 8.18 are implemented to analyze the outcome of We and Nb on the heat and mass transmission ratios.

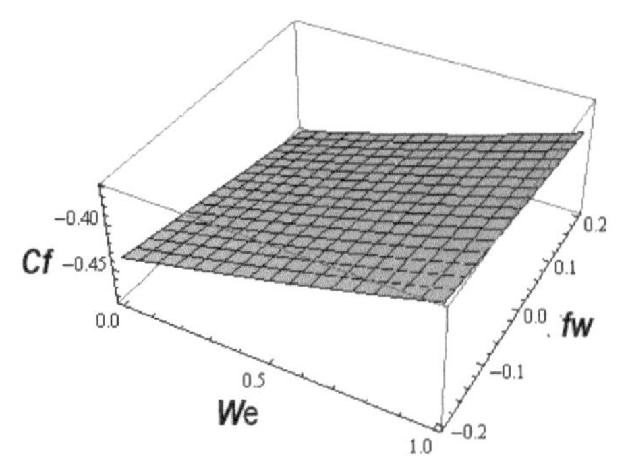

Figure 8.15 Impact of Cf for We and fw.

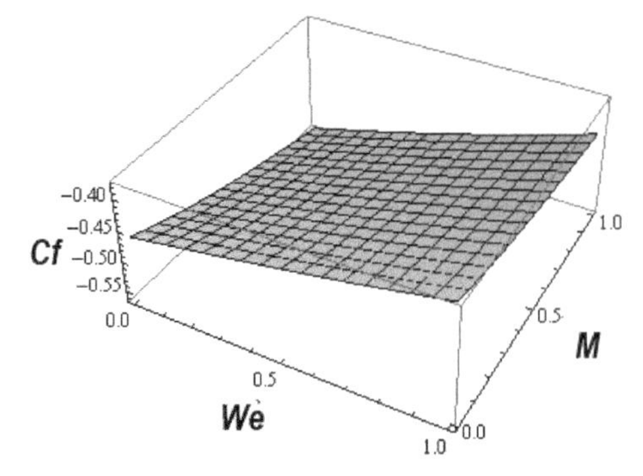

Figure 8.16 Impact of Cf for We and M.

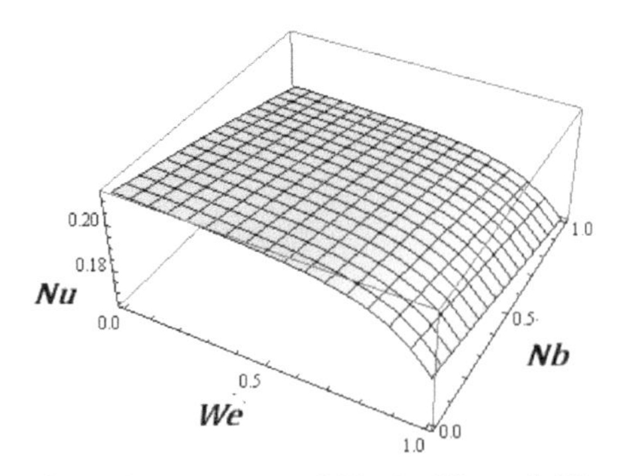

Figure 8.17 Impact of Nu for We and Nb.

We found that there is a decrement in heat transfer for higher values of We and Nb (see Figure 8.17). Figure 8.18 explores the mass transfer boosts substantially by improving We and Nb.

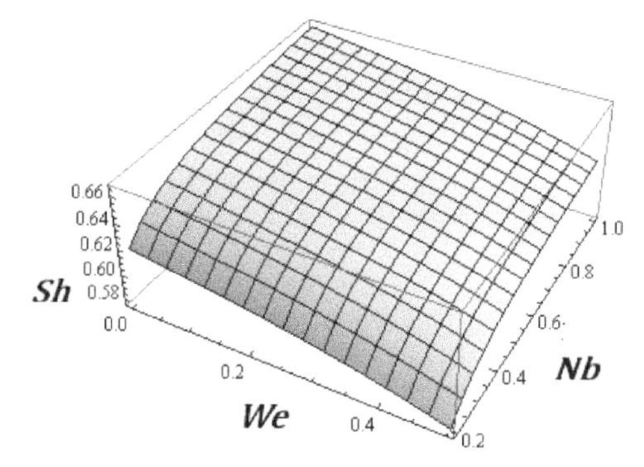

Figure 8.18 Impact of Sh for We and Nb.

8.6 Code Validation

A qualified validation is examined for skin friction is complete with Khan *et al.* [11] and Nadeem *et al.* [25] with the specific case of $M = 0$ (see Tables 8.2 and 8.3). Code validation of restricted Nusselt number and Sherwood number is found with formerly published references [26–29]. The validation is shown in Tables 8.4 and 8.5. An exceptional agreement is established between them. Therefore, the present computation is considered as an optimum one.

Table 8.2 Code validation of skin friction $(-Re^{0.5}Cf)$ for limiting case $M = 0$.

We	Nadeem [25]	Present
0.0	1.00000	1.00000
0.1	0.976588	0.97659
0.2	0.939817	0.93982
0.3	0.88272	0.88274

Table 8.3 Compression table for skin friction $(-Re^{0.5}Cf)$.

We	M	[11]	Present
0.1	0.1	2.05608	2.05608
0.2		2.01101	2.01101
0.3		1.96124	1.96124
0.1	0.2	2.14587	2.14587
	0.3	2.23184	2.23184

Table 8.4 Comparison values of the reduced Nusselt number with the special case $Rd = Ec = Nt = Nb = We = 0$, $Bi \to \infty$.

Pr	[26]	[27]	[28]	[29]	Present
0.20	0.1691	0.1691	0.1691	0.1691	0.1691
0.70	0.4539	0.5349	0.4539	0.4539	0.4539
2.00	0.9114	0.9114	0.9113	0.9114	0.9114
7.00	1.8954	1.8905	1.8954	1.8954	1.8954

Table 8.5 Comparison result of the reduced Nu and Sh for various values of Pr when $Nt = Nb = 0.5,\ Le = 5,\ Bi = 0.1,\ Rd = Ec = We = 0$.

Pr	[29]		Present	
	Nu	Sh	Nu	Sh
1	0.0789	1.5477	0.0789	1.5477
2	0.0806	1.5554	0.0806	1.5554
5	0.0735	1.5983	0.0735	1.5983

8.7 Key Results

The following main outcomes are given below:

1. Higher Weissenberg number reduces the velocity of the fluid, surface drag force, entropy generation, mass, and heat transfer rates.
2. Thermal relaxation time in entropy profile has been examined and introduced. The entropy of the system is found to be increased higher thermal relaxation time.
3. Entropy system is found to be increased for magnetic parameter and Brinkman number.

References

[1] M. M. Molla, L. S Yao. 2009. Mixed convection of non-Newtonian fluids along a heated vertical flat plate, Int. J. Heat Mass Transfer. 52: 3266–3271.

[2] K. Loganathan, S. Sivasankaran, M. Bhuvaneshwari, S. Rajan. 2019. Second-order slip, cross-diffusion and chemical reaction effects on magneto-convection of Oldroyd-B liquid using Cattaneo–Christov heat flux with convective heating, J Therm Anal Calorim. 136: 401–409.

[3] B. Sahoo. 2010. Flow and heat transfer of a non-Newtonian fluid past a stretching sheet with partial slip, Commun. Nonlinear Sci. Numer. Simul. 15: 602–615.

[4] M. Goyal, R. Bhargava. 2013. Numerical solution of MHD viscoelastic nanofluid flow over a stretching sheet with partial slip and heat source/sink, ISRN Nanotechnol. https://doi// 1155/2013/931021.
[5] Y. Khan, A. Hussain, N. Faraz. 2012. Unsteady linear viscoelastic fluid model over a stretching/shrinking sheet in the region of stagnation point flows, Sci. Iran. B. 19: 1541–1549.
[6] M.Y. Malik, I. Zehra, S. Nadeem. 2012. Numerical treatment of Jeffrey fluid with pressure-dependent viscosity, Int. J. Numer. Meth. Fluids, 68: 196–209.
[7] M. Irfan, M. Khan, W.A. Khan. 2019. Influence of binary chemical reaction with Arrhenius activation energy in MHD nonlinear radiative flow of unsteady Carreau nanofluid: dual solutions, Appl. Phys. A. https://doi.org/10.1007/s00339-019-2457-4.
[8] M. Turkyilmazoglu, I. Pop. 2013. Exact analytical solutions for the flow and heat transfer near the stagnation point on a stretching/ shrinking sheet in a Jeffrey fluid, Int. J. Heat Mass Transfer. 57: 82–88.
[9] K. Loganathan, S. Rajan. 2020. An entropy approach of Williamson nanofluid flow with Joule heating and zero nanoparticle mass flux, J. Therm Anal Calorim. 141: 2599–2612.
[10] B. Golafshan, A.B. Rahimi. 2018. Effects of radiation on mixed convection stagnation-point flow of MHD third grade nanofluid over a vertical stretching sheet. J Therm Anal Calorim. https://doi.org/10.1007/s10973-018-7075-4.
[11] M. I. Khan, S. Qayyum, T. Hayat, M. Imran Khan, A. Alsaedi, 2019. Entropy optimization in flow of Williamson nanofluid in the presence of chemical reaction and Joule heating, Int. J. Heat Mass Transfer, 133: 959–967.
[12] S. Baag, S. R. Mishra, G. C. Dash, M. R. Acharya. 2017. Entropy generation analysis for viscoelastic MHD flow over a stretching sheet embedded in a porous medium, Ain Shams Eng. J. 8: 623–632.
[13] N. Dalir. 2014. Numerical study of entropy generation for forced convection flow and heat transfer of a Jeffrey fluid over a stretching sheet, Alex. Eng. J. 53: 769–778.
[14] M. H. Abolbashari, N. Freidoonimehr, F. Nazari, M. M. Rashidi. 2015. Analytical modeling of entropy generation for Casson nano-fluid flow induced by a stretching surface. Adv. Powder Technol, 26: 542–552.
[15] A.S. Butt, A. Ali. 2014. Entropy analysis of magnetohydrodynamic flow and heat transfer due to a stretching cylinder, Journal of the Taiwan Institute of Chemical Engineers. 45: 780–786.

[16] A Bejan. 1979. A study of entropy generation in fundamental convective heat transfer, J. Heat Transfer. 101: 718–725.
[17] Y. Liu, Y. Jian, W. Tan. 2018. Entropy generation of electro magnetohydrodynamic (EMHD) flow in a curved rectangular micro channel, Int. J. Heat Mass Transf. 127: 901–913.
[18] C. Wang, M. Liu, Y. Zhao, Y. Qiao, Yan. J. 2018. Entropy generation analysis on a heat exchanger with different design and operation factors during transient processes, Energy. 158: 330–342.
[19] O. D Makinde, E. Osalusi, 2005. Second law analysis of laminar flow in a channel filled with saturated porous media. Entropy. 7: 148–160.
[20] M. M. Rashidi. and N. Freidoonimehr. 2014. Analysis of Entropy Generation in MHD Stagnation-Point Flow in Porous Media with Heat Transfer, International Journal for Computational Methods in Engineering Science and Mechanics. 15: 345–355.
[21] S. Liao, Y. A. Tan. 2007. general approach to obtain series solutions of nonlinear differential. Stud Appl Math. 119(5): 297–354.
[22] K. Loganathan, G. Muhiuddin, AM. Alanazi, FS. Alshammari, B M. Alqurashi and S. Rajan. 2020. Entropy Optimization of Third-Grade Nanofluid Slip Flow Embedded in a Porous Sheet With Zero Mass Flux and a Non-Fourier Heat Flux Model. Front. Phys. 8: 250.
[23] A. Aquino. I, Bo-ot, L. Ma. T. 2016. Multivalued behavior for a two-level system using Homotopy Analysis Method, Physica A. 443: 358–371.
[24] K. Loganathan, K. Mohana, M. Mohanraj, P. Sakthivel, S. Rajan. 2020. Impact of third-grade nanofluid flow across a convective surface in the presence of inclined Lorentz force: an approach to entropy optimization, J. Therm Anal Calorim, https://doi.org/10.1007/s10973-020-09751-3.
[25] S. Nadeem, S.T. Hussain, C. Lee. 2013. Flow of Williamson fluid over a stretching sheet, Braz. J. Chem. Eng. 30: 619–625.
[26] C. Y Wang. 1989. Free convection on a vertical stretching surface, ZAMM – J. Appl. Math. Mech./Z. Angew. Math. Mech. 69: 418–420.
[27] R Reddy Gorla, I. Sidawi. 1994. Free convection on a vertical stretching surface with suction and blowing, Appl. Sci. Res. 52: 247–257.
[28] W. A. Khan I. Pop. 2010. Boundary-layer flow of a nanofluid past a stretching sheet, Int. J. Heat Mass Transf. 53: 2477–2483.
[29] O. D. Makinde, A. Aziz, 2011. Boundary layer flow of a nanofluid past a stretching sheet with a convective boundary condition, Int. J. Therm. Sci. 50: 1326–1332.

9

Computational Study of Double Diffusive MHD Buoyancy Induced Free Convection in Porous Media with Chemical Reaction and Internal Heating

S. Kapoor

Regional Institute of Education (NCERT), Bhubaneswar,
Odisha 751022, India
E-mail: saurabh09.ncert@gmail.com

Abstract

This, chapter is a part engaged to build up a numerical model of the lightness incited convective stream and mass exchange of a two-dimensional, synthetically responding liquid over a vertical extending plane embedded in a non-Darcy permeable medium. The fluid with micropolar properties is taken into account for the entire study subjected to a uniform magnetic field. The combined effects of free convective heat and mass transfer are also studied in this chapter. The physical system is transformed into nonlinear system of PDE in the form of continuity, linear and angular momentum, energy, concentration equations, etc. Using a similarity transformation, the nonlinear coupled differential equations governing the boundary layer flow, heat, and mass transfer are first reduced to a system of coupled ordinary differential equations and then solved numerically using computational technique named finite-element technique. FORTRAN 77 package is used for the computational study. The impact of distinguished physical parameters such as magnetic field, Schmidt number (Sc), Grash of number, chemical reaction, etc., is performed in detail to create the interesting aspects of the solution. The implementation of numerical method is also validated with the published results to show the accuracy of the adopted method. In order to get a better understanding of the problem, the local Nusselt number and skin friction is also presented.

Keywords: Free convection, porous media, finite-element method (FEM), computation, chemical reaction.

9.1 Introduction

Fundamental and practical introduction to the use of computational methods, specially finite difference method, in the numerical simulation of fluid flows in porous media is of great interest to researchers, scientist, and the young generation. The variation of study covers a wide variety of flows, including single-phase, multiphase, volatile, non isothermal, and chemical compositional flows in both ordinary porous and fractured porous medium. In addition, a range of computational methods is used, and benchmark problems of nine comparative solution projects organized by the Society of Petroleum Engineers. The main aim of the present chapter is to highlight the broad spectrum of computation study. Computational methods for flows in porous media with chemical reaction show the flow equations and computational methods to introduce basic terminologies and notation. A study of boundary layer behavior on a stretching surface has attracted the attention of many researchers in the last decade as the analysis of such flows finds application in different areas such as aerodynamic extrusion of plastic sheets, the boundary layer along material handling conveyors, the cooling of an infinite metallic plate in a cool bath, and the boundary layer along a liquid film in condensation processes. In particular, the flow problems with heat and mass transfer over a stretching surface finds numerous industrial applications such as in the manufacture of sheeting material through an extrusion process and metallurgy where hydromagnetic techniques have been used. To be more specific, it may be pointed out that many metallurgical processes involve the cooling of continuous strips or filaments by drawing them through a quiescent fluid and that in the process of drawing, the strips are sometimes stretched. Due to these applications, the main part of this work deals with such kinds of problem. The study with convection in porous medium is well documented in the book of Nield [7], with compound response having various applications in numerous parts of designing science including hypersonic streamlined features [1, 2], geophysics and volcanic frameworks [3], reactant advancements [4], and synthetic designing cycles [5]. Numerous such examinations have been finished with limit layer hypothesis. Acrivos [6] examined the laminar limit layer stream with quick synthetic responses. Usman *et al.*, [10, 11] have brought up some angle while centering at unsteady MHD micropolar flow and mass transfer past a vertical permeable plate with variable suction and afterward endeavoring the issue while fusing the compound response boundary in radiation-convection flow in porous medium; yet, at the same time, numerous later investigations anyway did not consider the impact of substance response or species movement on the stream system. Sheikholeslami [12] concentrated logically and mathematically the liquid stream elements. In the current examination, we consider mathematically the lightness instigated convective stream and mass exchange of a micropolar, synthetically responding liquid over a vertical extending plane implanted in a DF permeable medium. The FEM has been used to settle the numerical model which establishes a two-point limit esteem issue. Such an examination finds significant applications in geochemical frameworks and furthermore substance reactor measure designing the above taken from the book [8, 9]. Bhargava *et al.* [21] analyzed the micropolar stream between turning circles. Ask *et al.* [13, 14] has considered and explored the warmth and mass exchange wonders in permeable media utilizing micropolar liquid, and afterward, they utilized computational limited component procedure for 2-Dl issue in channel. In this continuation, Rawatetal. [15, 16] have utilized the above procedure for the warmth and mass exchange wonders while consolidating the soret and duffor impacts and subsequently explored the thermophysical impacts utilizing MHD micropolar liquid in permeable media. The study is extended by Rawat and Kapoor [17] via studying the heat and mass transfer in a nonlinear stretching sheet using finite-element method (FEM) with micro inertia properties. Aharonov *et al.* [18] studied the three-dimensional reactive flow in porous media with dissolution effects. Later, Fogler and Fredd [19] analyzed the chemically reactive flow in porous media. In many industrial processes, engineers are primarily concerned with flow and transport phenomena over accelerating and stretching surfaces. In this regard, many studies have also been communicated. Sakiadis [20] first studied the laminar boundary layer flow past a continuous flat surface. The above initial study by Crane *et al.* [21] has examined the boundary layer flow caused by a stretching sheet with application in continuous casting, drawing of plastic films, and crystal growing. The same is extensively reported by Van Gorder *et al.* [22] in the case of nonlinear stretching sheet. The studies attract many researchers and scientists in this field. The theory and simulation of dynamics of micropolar fluid is well documented in the paper of James [23].

In continuation of the above work, Aurangzaib *et al.* [24] reported the flow of micropolar fluid and heat transfer over an exponentially permeable shrinking sheet. However, the theory of Navier-Stokes equations with adequate properties of micropolar fluid is nicely explained by Eringen *et al.* [25, 26]. Very recently, Shamshuddin and Thhuma [27] studied numerically the dissipation effect on inclined micropolar fluid flow with heat source and sink. The same studyat curved surface is extended by Yasmin [28]; he reported remarkable results in this direction.

Keeping in view the above literature, our objective is to understand the impact of different terminology for magneto-convection heat transfer of a two-phase, electrically conducting, particle suspension in a channel.

9.2 Mathematical Model

It is to be assumed that there is a steady-state, two-dimensional, laminar boundary layer flow with heat and solute transport properties. The fluid is assumed to be micropolar in nature. Here, it is to be considered that the fluid is passed through a vertical stretching surface, whereas the surface is embedded in a porous media. Here, the fluid is taken as having chemically dominating properties in the xy-coordinate system and the y-axis is perpendicular to the x-axis. It is assumed that the other properties of micropolar fluid are constant in the entire medium except the basic properties. In parametric study, the solute diffusion, fluid density, and viscosity are fixed. However, the fluid

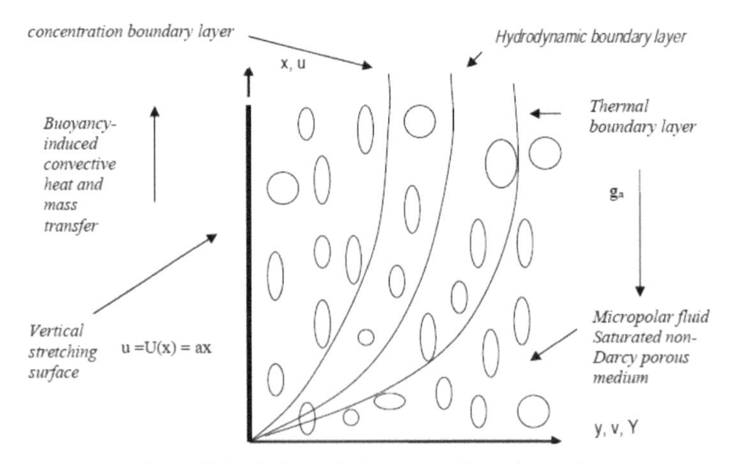

Figure 9.1 Schematic diagram of flow dynamics.

medium is assumed to be isotropic in nature. As the fluid is free convective, the free stream velocity, temperature, and concentration are almost negligible, i.e., almost 0. The internal heating is assumed to be effective at the boundary of the fluid flow regime. The governing boundary layer flow is modeled in the form of system of partial differential equation which includes the second-order Forchheimer drag force with Darcian effect. The local heat and solute transport is considered in intermediate stage of the flow dynamics with base velocity. Under the above assumption, the following system of equations is formed:

Equation of continuity:

$$\frac{\partial \mathrm{u}}{\partial \mathrm{x}}+\frac{\partial \mathrm{u}}{\partial \mathrm{y}}=\mathbf{0}. \tag{9.1}$$

Equation of momentum (linear):

$$u\frac{\partial \mathrm{u}}{\partial \mathrm{x}}+\mathrm{v}\frac{\partial \mathrm{u}}{\partial \mathrm{y}}=\frac{\mu}{+\mathrm{k}}\rho\left(\frac{\partial^2 \mathrm{u}}{\partial \mathrm{y}^2}\right)+\frac{\mathrm{k}}{\rho}\left(\frac{\partial \Omega}{\partial \mathrm{y}}\right)+\mathrm{g}[\beta_{\mathrm{T}}(\mathrm{T}-\mathrm{T}_\infty)$$

$$+\beta_{\mathrm{S}}(\mathrm{c}-\mathrm{c}_\infty)]-\frac{\mu}{\mathrm{k_p}}\mathrm{u}-\frac{\mathrm{b}}{\mathrm{k_p}}\mathrm{u}\,|\mathrm{u}|-\frac{\sigma_0 \mathrm{B}_0^2}{\rho}\mathrm{u}. \tag{9.2}$$

Equation of momentum (angular):

$$\mathrm{u}\frac{\partial \Omega}{\partial \mathrm{x}}+\mathrm{v}\frac{\partial \Omega}{\partial \mathrm{y}}=-\frac{\mathrm{k}}{\rho_{\mathrm{j}}}\left(2\Omega+\frac{\partial \mathrm{u}}{\partial \mathrm{y}}\right)+\frac{\mathrm{k}}{\rho_{\mathrm{j}}}\frac{\partial^2 \Omega}{\partial \mathrm{y}^2}. \tag{9.3}$$

Equation of energy:

$$\mathrm{u}\frac{\partial \mathrm{T}}{\partial \mathrm{x}}+\mathrm{v}\frac{\partial \mathrm{T}}{\partial \mathrm{y}}=\alpha\left(\frac{\partial^2 \mathrm{T}}{\partial \mathrm{y}^2}\right)+\mathbf{Q}_0(\mathrm{T}-\mathrm{T}_{\mathrm{w}}). \tag{9.4}$$

Equation of concentration or spices:

$$\mathrm{u}\frac{\partial \mathrm{c}}{\partial \mathrm{x}}+\mathrm{v}\frac{\partial \mathrm{c}}{\partial \mathrm{y}}=\varepsilon\left(\frac{\partial^2 \mathrm{c}}{\partial \mathrm{y}^2}\right)-\chi_e c. \tag{9.5}$$

With the boundary conditions:

$$\mathrm{u}=\mathrm{U}(\mathrm{x})=\mathrm{ax},\ \mathrm{v}=0,\ \mathrm{T}=\mathrm{T}_{\mathrm{w}},\ \mathrm{c}=\mathrm{c}_{\mathrm{w}}\&\boldsymbol{\Omega}=-\mathrm{s}\left(\frac{\partial \mathrm{u}}{\partial \mathrm{y}}\right)\ at\ \mathrm{y}=0$$

$$u=v=0,\quad \mathbf{T}=\mathbf{T}_\infty,\ \mathbf{c}=\infty\ \&\Omega\rightarrow 0\ at\ y\rightarrow\infty \tag{9.6}$$

where u and v are the velocity and Ω is the micro rotation velocity parameters. T is temperature and c is a concentration parameter. Q_0 is internal heating; apparent kinematic viscosity is denoted by $\frac{\mu+\mathrm{k}}{\rho}$ and $\frac{\mathrm{k}}{\rho}$ is a constant due to coupling. χ_{e} is the effective chemical reaction.

9.3 Similarity Transformation

In order to solve the above set of differential equation, it is necessary to non-dimensionalize the physical parameters and equations. The following set of linear transformation parameter is used to transform the dimensional parameters:

$$\psi = \left[\frac{\mu+k}{\rho}x\,U(x)\right]^{\frac{1}{2}},\ Y = \left[\frac{U(x)}{\frac{\mu+k}{\rho}}\right]^{\frac{1}{2}},\ u = \frac{\partial\psi}{\partial y}\ \text{and}\ v = -\frac{\partial\psi}{\partial x}$$

$$\Omega = \sqrt{\frac{U(x)}{\frac{\mu+k}{\rho}}}\,U(x)\,g(Y),\ U(x) \equiv ax,\ \theta = \frac{T-T_\infty}{T_w - T_\infty},\ C = \frac{c-c_\infty}{c_w-c_\infty}. \tag{9.7}$$

Using the above in Equations (9.1)–(9.5), the system of PDE is reduced to the set of ODE.

Non-dimensional equation of momentum (linear):

$$f''' + B_1 g' + ff'' + f^2 + Gr_x Re_x \theta + Gr_c Re_x C - \frac{1}{Re_x Da_x} f' - \frac{Fn_x}{Da_x} f^2 = 0. \tag{9.8}$$

Non-dimensional equation of momentum (angular):

$$\lambda g'' - \frac{\lambda}{\Lambda}\left(2g + f''\right) - f'g + g'f = 0. \tag{9.9}$$

Non-dimensional equation of temperature:

$$\theta'' + Prf\theta' + Q\theta = 0.] \tag{9.10}$$

Non-dimensional equation of concentration:

$$C'' + ScfC' - Sc\left[\chi Re_x C + Cf'\right] = 0 \tag{9.11}$$

where

$$Re_x = \frac{Ux}{\frac{\mu+k}{\rho}},\ Gr_c = \frac{\frac{\mu+k}{\rho} g\beta_S (c - c_\infty)}{U^3},\ Gr_x = \frac{\frac{\mu+k}{\rho} g\beta_T (T - T_\infty)}{U^3}$$

$$Da_x = \frac{k_p}{x^2},\ Fn_x = \frac{b}{x},\ Pr = \frac{\frac{\mu+k}{\rho}}{\alpha},\ Sc = \frac{\frac{\mu+k}{\rho}}{\varepsilon},\ \chi = \frac{\chi e\frac{\mu+k}{\rho}}{U^3}. \tag{9.12}$$

The corresponding Boundary conditions:

$$f(0) = 0,\ f'(0) = 1,\ \theta(0) = 1,\ C(0) = 1,\ g(0) = -sf''(0)\ \text{at } Y = 0$$
$$f' \to 0,\ \theta \to 0,\ C \to 0,\ g \to 0,\ \text{at } Y \to \infty \tag{9.13}$$

where as the local heat transfer coefficient (Nusselt number) is defined as follows:

$$Nu_x = \frac{q_w}{(\mathbf{T_w} - \mathbf{T_\infty})} \frac{x}{k} = -(Re_x)^{\frac{1}{2}} \frac{\partial \theta}{\partial Y}(0). \tag{9.14}$$

9.4 Computational Solution

The set of equation from (9.8) to (9.11) along with the boundary condition (9.13) is solved using the computational method. The implicit finite difference technique is used to find the solution of the above set of differential equations:

$$f''' = \frac{f(i+2) - 2f(i+1) - f(i-2) + 2f(i-1)}{2h^3} + o(h^2)$$
$$f'' = \frac{f(i+1) - f(i) + f(i-1)}{h^2} + o(h^2)$$
$$g'' = \frac{g(i+1) - g(i) + g(i-1)}{h^2} + o(h^2)$$
$$\theta'' = \frac{\theta(i+1) - \theta(i) + \theta(i-1)}{h^2} + o(h^2)$$
$$C'' = \frac{C(i+1) - C(i) + C(i-1)}{h^2} + o(h^2)$$
$$f' = \frac{f(i+1) - f(i-1)}{2h} + o(h^2)$$
$$g' = \frac{g(i+1) - g(i-1)}{2h} + o(h^2)$$
$$\theta' = \frac{\theta(i+1) - \theta(i-1)}{2h} + o(h^2)$$
$$C' = \frac{C(i+1) - C(i-1)}{2h} + o(h^2) \tag{9.15}$$

Here, we have chosen step size $h = 0.0001$ (h is uniform grid spacing) to satisfy the convergence criterion of 10^{-7}. In all the cases, for the code validation, we used 10,000 grid point to maintain the accuracy of the solution.

Using Equation (9.15), the set of differential equation will become

$$\frac{f(i+2)-2f(i+1)-f(i-2)+2f(i-1)}{2h^3}+g\frac{g(i+1)-g(i-1)}{2h}$$
$$+f(i)\frac{f(i+1)-f(i)+f(i-1)}{h^2}+(f(i))^2+Gr_xRe_x\theta(i)$$
$$+Gr_cRe_xC(i)-\frac{1}{Re_xDa_x}\frac{f(i+1)-f(i-1)}{2h}-\frac{Fn_x}{Da_x}(f(i))^2=0 \tag{9.16}$$

$$\lambda\frac{g(i+1)-g(i)+g(i-1)}{h^2}-\frac{\lambda}{\Lambda}\left(2g(i)+\frac{f(i+1)-f(i)+f(i-1)}{h^2}\right)$$
$$-\frac{f(i+1)-f(i-1)}{2h}g+\frac{g(i+1)-g(i-1)}{2h}f=0 \tag{9.17}$$

$$\frac{\theta(i+1)-\theta(i)+\theta(i-1)}{h^2}++Prf\frac{\theta(i+1)-\theta(i-1)}{2h}+Q\theta(i)=0 \tag{9.18}$$

$$\frac{C(i+1)-C(i)+C(i-1)}{h^2}++Scf(i)\frac{C(i+1)-C(i-1)}{2h}$$
$$-Sc\left[\chi Re_xC(i)+C(i)\frac{f(i+1)-f(i-1)}{2h}\right]=0. \tag{9.19}$$

With the appropriate B.C. at initial guess given at (9.13). The linearized system of equation is solved using Gaussiterative method.

9.5 Computational Results and Their Interpretation

In order to explain the physics behind the physical model the comparison is made for the computational study with existing results in some special case. Table 9.1 is given to compare results with that of Rawat *et al.*

The table is drawn for the distinguished value of physical parameters while omitting the effect of internal heat source and magnetic effect. It has been observed from the table that the result is exact upto five decimal places with published results; this leads to correctness of applied method or technique.

In order to discuss the validity of physical parameters, the impact of distinguished parameter is shown graphically in this section. Figure 9.2 shows the effect of media permeability in order to see the strength of fluid flow passes through porous medium. The variation in velocity is observed while fixing other parameters.

Table 9.1 Comparison with the published results at $Q = 0$.

Y	Rawat *et al.*	Present for Free Free Stream Velocity	Rawat *et al.*	Present for Angular Velocity
0	1	1	0.259095	0.259012
0.8	0.673322	0.673285	0.148205	0.148164
1.6	0.453918	0.453893	0.093566	0.093535
2.4	0.297017	0.297001	0.061379	0.061357
3.2	0.191239	0.191227	0.040084	0.040072
4	0.121795	0.121791	0.026326	0.026319
4.8	0.075524	0.075505	0.017486	0.017467
5.6	0.043871	0.043843	0.011467	0.011436
6.4	0.022071	0.022031	0.006937	0.006902
7.2	0.007742	0.007699	0.003192	0.003145
8	0	0	0	0

Table 9.2 Comparison with the published results at $Q = 0$.

χ	Rawat *et al.* $\theta'(0)$	Present study $\theta'(0)$
	2	3
0	0.347135	0.347138
1	0.334291	0.334293
5	0.314796	0.314798
10	0.306021	0.306022
20	0.198679	0.298698
1.		4

It has been observed from the figure that the velocity decreases while decreasing the media permeability. The velocity is asymptotically stable near the boundary layer. The fluid flow also shows an asymptotical behavior. The above is observed for the fixed value of $\mathrm{Pr} = 0.7$, $\mathrm{Sc} = 0.1$, $s = 0.5$, $\mathrm{Re}_x = 1$, $\mathrm{Fn}_x = 1$, $\mathrm{Gr}_x = 1$, $Gc_x = 1$, $\chi = 1$, $M = 1$, $Q = 1$, and B = 0.02. The value of Darcy number is taken from 0.1 to 10. The higher value of Da shows a negligible effect of porous medium. This is valid for the numerical study and not for the physical study. As in the case of the Forchheimer drag, there is no meaning of low permeability.

Figure 9.2 shows the effect of chemically reacting parameter χ while fixing the other parameters $\mathrm{Pr} = 0.7$, $\mathrm{Sc} = 0.1$, $s = 0.5$, $\mathrm{Re}_x = 1$, $\mathrm{Fn}_x = 1$, $Gr_x = 1$, $Gc_x = 1$, $\mathrm{Da} = 0.1$, $M = 1$, $Q = 1$, and $B = 0.02$. It has been observed from the figure that the velocity decreases while increasing

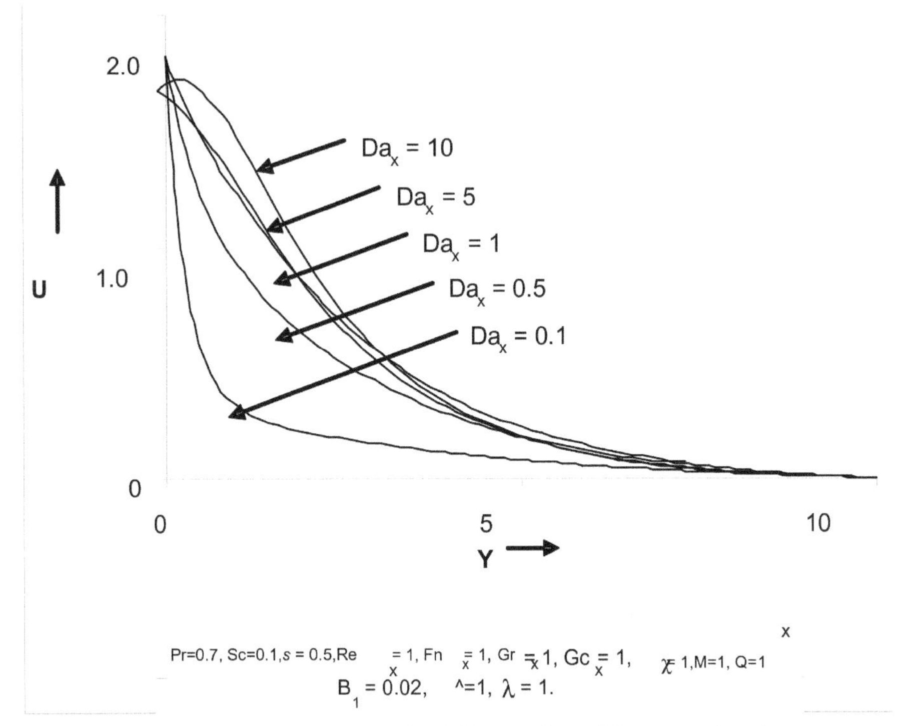

Figure 9.2 Velocity profile for different values of Darcy number Da.

the chemical reaction parameter. The velocity is asymptotically stable and flat near the boundary wall. Here the chemically reacting parameter χ is varied between 0 and 10^3. From the figure, it is also observed that high value of chemical parameter remains the same. The velocity profile is almost overlapped in entire values between 0 and 10 at characteristic length *Y*.

In order to understand the temperature and concentration variation with reference to the distinguished physical parameters, here, the effect of some of the physical parameter is shown.

In Figure 9.3, the temperature profile is plotted for the different values of Prandtl number Pr for the fixed values of other parameter like $Sc = 0.1$, $s = 0.5$, $Re_x = 1$, $Fn_x = 1$, $Gr_x = 1$, $Gc_x = 1$, $\chi = 1$, $M = 1, Q = 1$, and $B = 0.02$. The basic phenomenon of Pr is that the **Prandtl number** controls the relative thickness of the momentum and thermal boundary layers. However, when Pr is small, it means that the heat diffuses quickly compared to the velocity (momentum); it is the ratio of momentum diffusivity to the thermal diffusivity. Here, it can be seen from the profile that as we increased

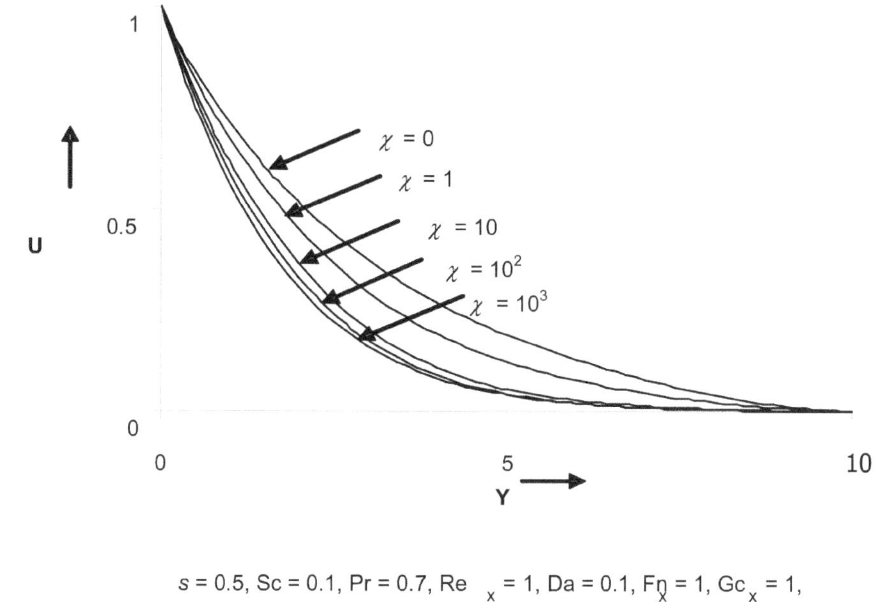

Figure 9.3 Velocity profile for different values of chemical reaction parameter χ.

the value of Prandtl number, the heat during the boundary layer is decreased. Here, it is also important to note how temperature profile decreases within the range of 0.25 and 0.5. It can be seen from the figure that the temperature decreases from the plate to boundary for all the values of Pr. The typical behavior of temperature shows that heat is reduced near the boundary wall. The structure of pores also allows the heat absorption.

Another important point to be noted is that the variation in profile also asymptotically converges from the wall to boundary. The characteristic length of the wall is fixed at 10, which is only for the numerical computation. In the physical model, it can be noted that the boundary may be large enough.

In order to investigate the solute transport phenomena, Figure 9.4 is plotted for the different value of Sc. It is to be seen from the figure how the Schmidt number affects the solute profile. The Schmidt number is the ratio of the shear component for diffusivity viscosity/density to the diffusivity for mass transfer D. It physically relates the relative thickness of the hydrodynamic layer and mass-transfer boundary layer. The profile of C is shown for fixing the value of $Pr = 0.7$, $s = 0.5$, $Re_x = 1$, $Fn_x = 1$, $Gr_x = 1, Gc_x = 1$, $\chi = 1$, $M = 1$, $Q = 1$, and $B = 0.02$. The impact

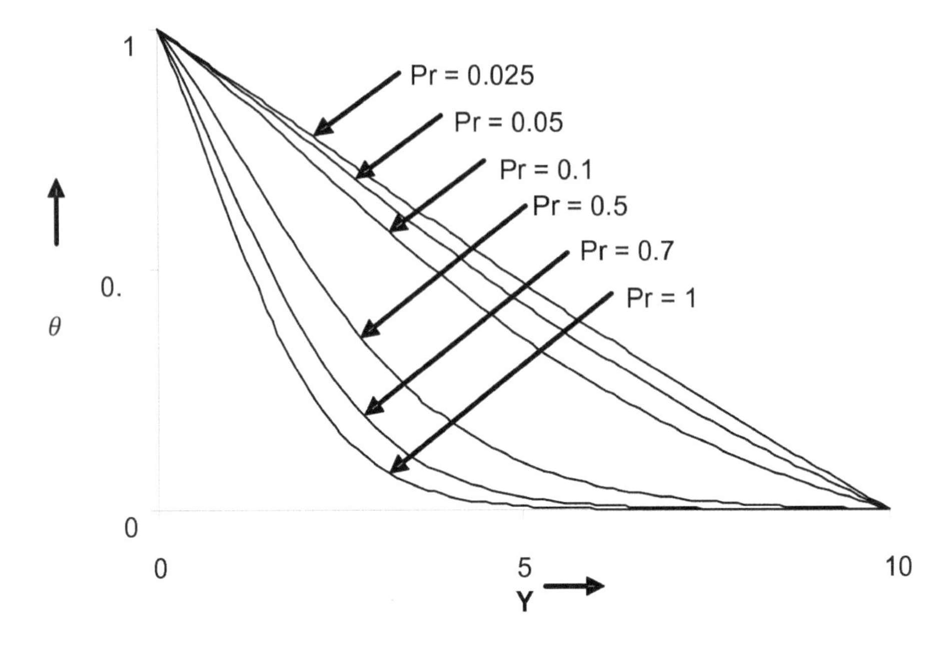

$s = 0.5$, $Sc = 0.1$, $Re_x = 1$, $Da_x = 0.1$, $Fr_x = 1$, $Gr_x = 1$, $Gc_x = 1$,
$_1 = 0.02$, $\lambda =$ $\chi = M = Q = 1$.

Figure 9.4 Temperature profile for different values of Pr.

of Schmidt number (Sc) on the mass exchange work is outlined for both the responsive stream case and the non-receptive stream case. Here, it can be seen that Sc measures the general viability of force and species movement by dispersion. More modest Sc esteems can speak to, for instance, hydrogen gas as the species diffuse (Sc = 0.1 to 0.2). Sc = 1.0 compares roughly to carbon dioxide diffusing in air; Sc = 2.0 suggests Benzene diffusing in air and higher qualities to oil subordinates diffusing in air (for example, Ethylbenzene) as demonstrated by Gebhart. Calculations have been performed for Pr = 0.7; so Pr is not equal to Sc. Actually, this infers that the warm and species dissemination areas are of various degrees. As Sc increments, for the receptive stream case, concentration emphatically diminishes since bigger estimations of Sc are identical to a decrease in the synthetic sub-atomic diffusivity; for example, less dispersion in this manner happens by mass vehicle. All profiles are believed to slide from a greatest grouping of 1 at $Y = 0$ (the wall) to zero.

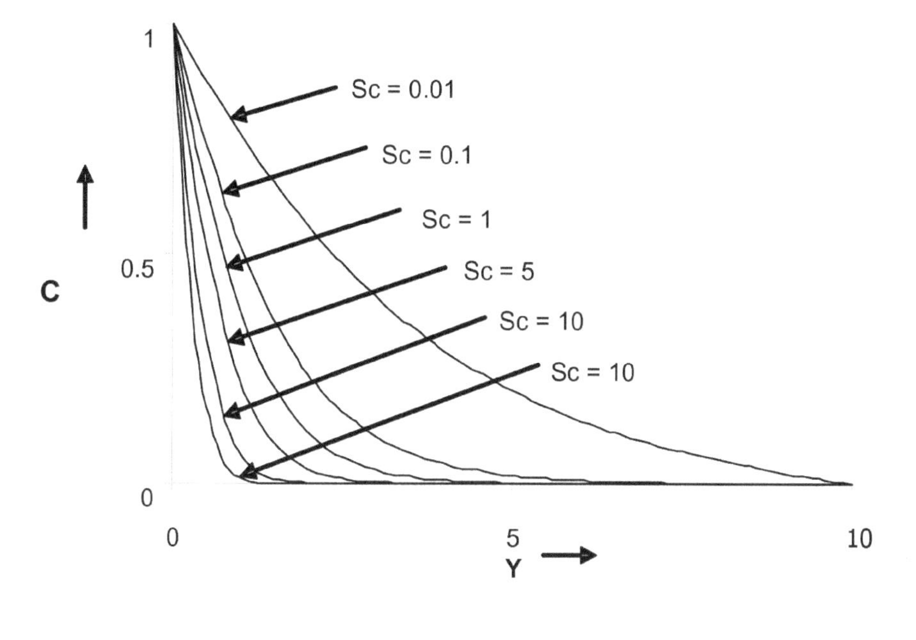

$s = 0.5$, $Pr = 0.7$, $Re_x = 1$, $Da_x = 0.1$, $Fr_x = 1$, $Gr_x = 1$, $Gc_x = 1$,
$B_1 = 0.02$, $M = 1$, $\lambda = 1$, $\chi = Q = 1$.

Figure 9.5 Concentration profile for different value of Schmidt number Sc.

9.6 Conclusion

The double-diffusive convection in a micropolar fluid in a vertical porous system is studied using computational simulation. The modified Darcy equation named Forchheimer drag includes the inertia term of the momentum equation. The similarity transformation is successfully used to non-dimensionalize the physical system. The system of differential obtained a function of the governing parameters. The effect of chemical reaction, Schmidt number, etc. The following conclusions are drawn:

1. The finite difference scheme is implemented successfully with an order of accuracy of order 10^{-7}.
2. The computational method gives faster and better results. The comparison is also shown in the tables.
3. The velocity, temperature, and concentration profiles show the effectiveness of the physical parameter.
4. The heat and solute phenomena show comparative results.

References

[1] Bertolazzi E., "A finite volume scheme for two-dimensional chemically-reactive hypersonic flow," Int. J. Num. Meth. Heat Fluid Flow, 8(8) (1998) 888–933.

[2] Kelemen P., Dick P. and Quick J., "Production of harzburgite by pervasive melt rock-reaction in the upper mantle," Nature, 358 (1992) 635–641.

[3] Zeiser T., Lammers P., Klemm E., Li Y.W., Bernsdorf J. and Brenner G., CFD Calculation of flow, dispersion and reaction in a catalyst filled tube by lattice Boltzmann method, Chem. Eng. Sci., 56 (4) (2001) 1697–1704.

[4] Levenspiel O., Chemical Reaction Engineering, John Wiley, New York, 3rd edition (1999).

[5] Acrivos A., "On laminar boundary layer flows with a rapid homogenous chemical reaction," Chem. Eng. Sci. 13 (1960) 57.

[6] Takhar H.S. and Soundalgekar V.M., On the diffusion of a chemically reactive species in a laminar boundary layer flow past a porous plate. "L" Aerotechnica Missili E. Spazio. 58 (1980) 89–92.

[7] D.A. Nield., Bejan., Convection in porous medium, Springer enlarge. (2006).

[8] K., Eriksson, D. Estep, P. Hansbo & C. Johnson, *Computational Differential Equations*, (Cambridge: Cambridge, UK), 1996.

[9] M.S. Gockenbach, *Partial Differential Equations: Analytical and Numerical Methods*, (SIAM: Philadelphia), 2002.

[10] H. Usman M. M. Hamza B.Y Isah., "Unsteady MHD Micropolar Flow and Mass Transfer Past a Vertical Permeable Plate with Variable Suction", International Journal of Computer Applications 36(4), 12–17, 2011.

[11] H. Usman M. M. Hamza M.O Ibrahim., "Radiation-Convection Flow in Porous Medium with Chemical Reaction", International Journal of Computer Applications 36(2), 12–16, 2011.

[12] M. Sheikholeslami., and D. D. Ganji. "Analytical investigation for Lorentz forces effect on nanofluid Marangoni boundary layer hydrothermal behavior using HAM." Indian Journal of Physics 91, no. 12 (2017): 1581–1587.

[13] O.A Bég., R. Bhargava., S. Rawat., H.S Takhar, T.A. Beg. "A Study of Steady Buoyancy- Driven Dissipative Micropolar Free Convective Heat and Mass Transfer in a Darcian Porous Regime with Chemical

Reaction," Nonlinear Analysis: Modeling and Control, 12, 2, 157–180 (2007).

[14] O.A Bég., R. Bhargava., S. Rawat., Kalim Halim and H.S. Takhar. "Computational modeling of biomegnatics micropolar blood flow and heat transfer in a two dimensional non-darcian porous channel." Meccanica 43: 391–410 (2008).

[15] S. Rawat and R. Bhargava. "Finite element study of natural convection heat and mass transfer in a micropolar fluid-saturated porous regime with soret/dufour effects," Int. J. of Appl. Math and Mech. 5, 2, 58–71. (2009).

[16] S. Rawat and R. Bhargava and O. Anwar Bég., "Hydromagnetic micropolar free convection heat and mass transfer in a darcy-forchheimer porous medium with thermo physical effects: finite element solutions," Int. J. of Appl. Math and Mech, 6, 13, 72–93 (2010).

[17] S. Rawat. S. Kapoor and R. Bhargava., "MHD flow heat and mass transfer of micropolar fluid over a nonlinear sheet with variable micro inertia density, heat flux and chemical reaction in a non darcy porous medium" J. of Applied Fluid Mechanics, 9(1), 321–331, 2016.

[18] Aharonov E., Spiegelman M. and Kelemen P., "Three-dimensional flow and reaction in porous media: implications for the earth's mantle and sedimentary basins", J. Geophys. Res. 102 (1997) 14821–14834.

[19] Fogler H.S. and Fredd C., "The influence of transport and reaction on wormhole formation in porous media", A I Chem E J. 44 (1998) 1933.

[20] Sakiadis B.C., Boundary layer behaviour on continuous solid surface II: "The boundary layer on a continuous flat surface", A I Chem E J. 7 (1961b) 221–225.

[21] L. J. Crane, "Flow past a stretching plate," *Zeitschrift für angewandte Mathematik und Physik*, vol. 21, no. 4, pp. 645–647, 1970.

[22] R. A. Van Gorder, K. Vajravelu, and F. T. Akyildiz, "Existence and uniqueness results for a nonlinear differential equation arising in viscous flow over a nonlinearly stretching sheet," *Applied Mathematics Letters*, vol. 24, no. 2, pp. 238–242, 2011.

[23] C. James, C.C. Liang and J.d. Lee, "Theory and simulation of micropolar fluid dynamics," Proceedings of the Institution of Mechanical Engineers Part N Journal of Nanoengineering and Nanosystems 224(1–2): 31–39, 2011.

[24] Md Aurangzaib, S. Uddin, K. Bhattacharya, S. Shafie, "Micropolar fluid flow and heat transfer over an exponentially permeable shrinking sheet," 5(4), pp. 310–317, 2016.

[25] A. C. Eringen, “Theory of micropolar fluids,” vol. 16, pp. 1–18, 1966.
[26] A. C. Eringen, “Theory of thermo microfluids,” *Journal of Mathematical Analysis and Applications*, vol. 38, no. 2, pp. 480–496, 1972.
[27] M.D. Shamhuddin and T. Thhuma, “Numerical study of a dissipative micropolar fluid flow past an inclined porous plate with heat source/sink.” Propulsion and Power Research, 8(1), pp. 56–68, 2019.
[28] A. Yasmin., K. Ali., M. Ashraf., “Study of heat and mass transfer in MHD micropolar fluid over a curved stretching sheet, Scientific report,” PMC 7067796, 2020.

10

Importance of Analytic Continuation in Complex Analysis

Kamna Singh and Geeta Arora

Lovely Professional University, Phagwara, India
E-mail: kamnakamnasingh13@gmail.com; geetadma@gmail.com

Abstract

Analyticity, of functions plays an important role in complex analysis, a branch of mathematics, because it explains the behavior of the function and is helpful in solving problems by using its properties. But when there exist different singularities in a function, then it is not easy to extend the radius of analyticity of the given function. In such a situation, the concept of analytic continuation plays an important role in extending the domain of analyticity which helps in knowing more about the behavior of the function. This is useful in understanding many of the real-life problems using the concept of homomorphic functions.

This chapter explores the basic and important concepts in the theory of analytic continuation with discussion on some of its advancement and applications also.

Keywords: Analytic continuation, singularities, power series, branch cuts, germ, sheaf, Riemann surfaces, Atlas, Complete analytic function, Complex chart, Direct analytic continuation, Indirect analytic continuation, Transition function.

10.1 Introduction

When Weierstrass, the German Mathematician, meticulously started to work on the theory of analytic function, he chose power series as the building blocks of his whole theory. He conscientiously differs the idea of his proposition from Riemann as he endorsed his outlook in a more geometric manner. Riemann strongly espouses the idea of singularities that analytic functions can be described by its singularities and some properties, but the radius of analyticity cannot be extended in case of different singularities. In the forward direction with the power series, after analyzing it thoroughly, it was seen that it has many limitations and challenges. This leads to the theory of "***Analytic Continuation***" that has proven as a powerful tool to overcome some of the limitations which were faced. Rigorize, complex analysis is based on the concept of power series because of convergence properties. Due to these convergence properties, the power series can be differentiated or integrated term by term, thus supplying a tool of immense importance. But along with advantages, it has limitations as all the functions cannot be represented by a single power series. This limitation can be overcome by the method of "**analytic continuation**", which is employed to extend the domain of a complex function for known suitable conditions.

Generally, most of the problems in applied science can be expressed in a form of mathematical equations and functions. A lot of valuable information can be extracted from the analysis of functions and the concern is about the extension of the domain in which required information exists. Here comes the need of analytic continuation as some of the functions require extension to the whole complex plane for the analysis of the required result.

As we know, analytic functions can be determined by many ways, for example, convergent power series, path integrals of continuous functions, etc., and in this regard, there arises two important questions. *Whether we can extend the domain of analytic functions larger than the given domain and why do we need it?* ***Yes***, we can. As whatever the way we are using to determine our analytic function, we do not know about the defining of our function outside the given domain; so to know how far we can define our analytic function or to know the largest domain in which the function is defined, we do the extension of our given domain which is known as analytic continuation.

It is found that those extensions will not always be unique.

Concept of analytic continuation defeats the limitations of power series and there exists a thread which leads to the "Riemann Surfaces", and analytic continuation is such a concept which is very much necessary to understand Riemann hypothesis.

The theory of Weierstrass has its intrigue entirely historical in consideration to the limitations of the power series and the convergence of their domains which is forsooth a hindrance rather than an aid. It must be taken into account that the idea of Weierstrass is still the foundation of the understanding regarding the multivalent values with the emphasis on the complex analytic function. One of the more consistent and rapid procedures that exist in several complex concepts that influence recent analytical theories and occur in the work related to many complex variables can be discussed with the concept of **germ and sheaf**.

Let us consider an ordered pair (f, K) where K is a point and f is a function analytic at point K. Here, f is a defined function and also analytic in some open set that contains K. Two pairs (f_1, K_1) and (f_2, K_2) shall be equivalent if and only if $K_1 = K_2$ and $f_1 = f_2$ in some neighborhood of K_1. The conditions for an equivalence relation are obviously fulfilled. The above-defined equivalence classes are called germs or, more specifically, germs of analytic functions and the set of all the germs is known as sheaf. It is to be known that any complex function which is differentiable has a depiction of local power series.

In 1831, the validation of *Taylor series expansion* was first proved by Cauchy, but, in 1715, *Brooke Taylor* was the first person who has communicated and published the idea that it is possible to expand a function as a power series in the form of

$$f(x+h) = \sum_{n=0}^{\infty} \frac{f^{(n)}(x)}{n!} h^n$$

The series got its name after him as the Taylor series. His theory was confined to the functions of reals only, and it is not surprising that this idea was widely known to others so far. Around 1800, there were numerous efforts to use power series as the plinth of the theory of real analysis, and one of the famous attempts was done by Joseph-Louis Lagrange, in 1797.

The use of power series in complex analysis is extensively done by Cauchy. In 1829, fortunately, he found a counter example $f(x) = e^{-1/x^\wedge 2}$ and with the help of this example, he showed that every infinitely differential real function is not equal to its Taylor series. After few years, he showed that valid power series expansions can be found for all the differentiable complex functions.

Cauchy illustrated that, in the case of reals, there are some functions which are infinitely differentiable, but those functions are not analytic. But in the case of complex, he showed that if a function is *once* differentiable in a domain, it must be analytic. In this way, he tries to show that complex analysis will be easier than the real analysis and turn down the common study of differentiable complex functions to calculate it with the power series.

A complex function is differentiable on a domain iff it is analytic on that domain.

The Taylor series expansions were seen to be restricted for various applications. Then in 1843, Laurent gives some generalization which was found to be useful. Laurent includes the *negative* powers in addition to the positive powers in power series. Let us understand the benefit of doing this with the following example, $f(z) = e^{-1/x\wedge 2}$; with respect to the Taylor series expansion, this function *f(z)* behaves very inadequately. It was observed that, about the origin bounded to the real line, Taylor series expansion would look like $0 + 0\mathrm{x} + 0\mathrm{x}^2 + \cdots$, and it will not converge to *f(x)*. On replacing z by $-1/z^2$ in e^z, the series representation will look like

$$f(z) = 1 - z^{-2} + 1/2!z^{-4} + 1/3!z^{-6} + \cdots$$

The above series is a kind of negative power series and it converges for all the z for which $-1/z^2$ is defined and $z \neq 0$.

As discussed earlier, the applications of analytic continuation are widely spread especially in the field of mathematics and physics. To define a function, a very general way is to do complex analysis by first taking the function into a small domain and then starting the extension. The *Riemann zeta function* and *Gamma function* are some of the examples of practicing this continuation. One of the applications of analytic continuation is that it develops the idea of Reimann surfaces via the idea of determining the analytic continuation of the given function. Riemann's idea was to replace C with intricated "Riemann surface". In a layman manner, it is to paste together the domains of the given function elements at wherever the function satisfies.

10.2 Analytic Continuation

As discussed earlier, we have acknowledged with the idea of analytic continuation, as it is a process used for the extension of the domain in the view of a given analytic function.

10.2.1 Direct Analytic Continuation

Direct analytic continuation is one of the simplest types of analytic continuation in which the domain of the given function is extended from one domain to another overlapping domain, larger than the previous one. The repetition of this method is considered as indirect analytic continuation.

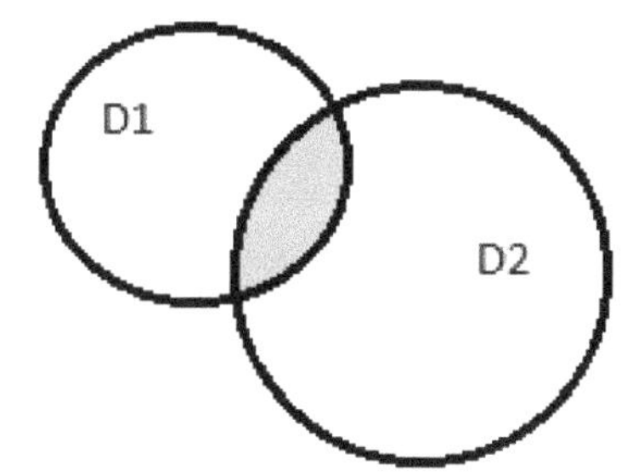

Figure 10.1 (a) Direct analytic continuation.

Definition 10.1. *If f_1 is analytic on a domain D_1 and f_2 is analytic on a domain D_2, where $D_1 \cap D_2 \neq \phi$ and $f_1(z) = f_2(z)$ for all $z \in D_1 \cap D_2$, then f_2 is a direct analytic continuation of f_1 to the domain D_2.*

Let us consider $f_1(z)$ to be a holomorphic function of z which is defined on the region D_1 in the complex plane. Let us consider another region D_2 in the complex plane such that the intersection of both the regions should not be empty, i.e., $D_1 \cap D_2 \neq \phi$. In the region D_2, a holomorphic function $f_2(z)$ is defined which coexist with holomorphic function $f_1(z)$ on the common region of the domains D_1 and D_2 which is known as the analytic continuation of $f_1(z)$ into D_1.

Here comes a natural question, *is it always possible to extend the domain of a given analytic function?* Obviously, ***No***. For example,

$$f(z) = 1/z, z \in D, \quad \text{where } D = \mathbb{C}\backslash\{0\}.$$

Here, the $f(z)$ does not have an extension to $\mathbb{C}$.

This shows that the extension of the domain of every given analytic function is not possible, but if it possibly exists, then that analytic continuation must be *unique* (by the following lemma).

Lemma: *Let f, g be analytic on a domain D. Suppose that P and Q are open sets and $\phi \neq P \cap Q \subseteq D$. Suppose that $p(z) = q(z)$ for all $z \in P \cap Q$, and f, g are analytic functions defined on D such that $f(z) = p(z)$ for $z \in P$ and $g(z) = q(z)$ for $z \in Q$. Then $f(z) = g(z)$ for all $z \in D$.*

In continuation of the above discussion, $f_1(z)$ is analytically continuable into D_1. If $z \in D_1$ and $z \in D_2$, putting $f(z) = f_1(z)$ and $f(z) = f_2(z)$, this leads to the extension of the holomorphic function $f(z)$ which is defined on the region $D_1 \cup D_2$ and the function $f(z)$ is known as the analytic continuation of $f_1(z)$ into D.

As the analytic extension f_2(z) of f_1 (z) into the region D_2 exists and similarly the analytic extension f_3(z) of f_2(z) into D_3 exists, then f_3(z) is known to be the analytic extension or the analytic continuation of f_1(z) into the region D_3.

Therefore, if we consider the given sequence of the regions,

$$\boldsymbol{D_1, D_2, D_3, \ldots, D_{n-1}, Dn.} \textbf{ where, } \boldsymbol{D_{k-1} \in D_k \neq \phi, k = 1, 2, 3, \ldots, n.}$$

And f_2(z) is the analytic continuation of f_1(z) into the region D_2, f_3(z) is the analytic continuation of f_2(z) into D_3,$\ldots$, f_n(z) is the analytic continuation of f_{n-1}(z) into the region D_n.

Thus, it can be concluded that the f_1(z) is analytically continuable across the sequence D_1, D_2, D_3,$\ldots$,D_{n-1}, D_n and all the functions $f_k(z)$ are known as the analytic continuations of f_1(z), where $k = 2, 3\ldots, n$.

These functions f_k(z) are the analytic continuation which are formed uniquely by f_1(z), and the regions D_1, D_2,$\ldots$, D_n, respectively. And as the functions f_2(z), f_3(z), f_4(z),$\ldots$,f_n(z) exists and are analytic continuations of f_1(z), then the defining of the holomorphic function f(z) will be on the region when we put f(z) $= f_k$(z) if z $\in D_k$.

$$D = D_1 \cup D_2 \cup D_3 \cup \ldots \cup D_n.$$

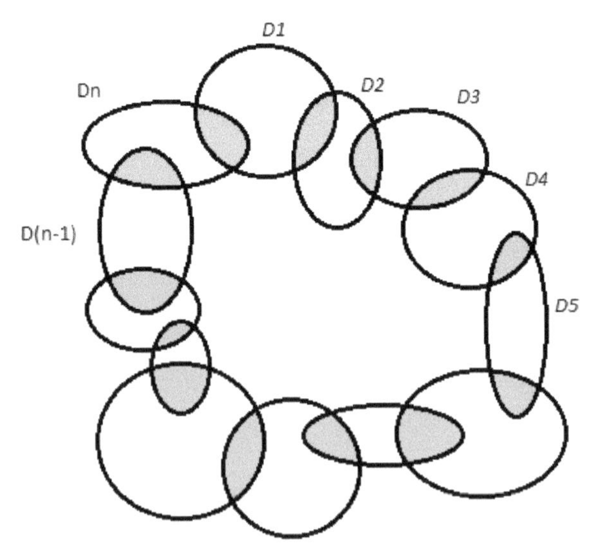

Figure 10.1 (b) Representation of Analytic Continuation into the given sequence of regions D_n.

10.2.2 Indirect Analytic Continuation

Suppose we have pairs (D_α, f_α) where D_α are domains and f_α: $D_\alpha \rightarrow C$ are analytic and a total ordering of a set $A = \{\alpha\}$. So, that means, there is a collection of pairs (D_α, f_α) indexed by the α running in the indexed set A and each of these pairs consists of domain D_α which is namely an open connected set and an analytic function f_α on the given domain.

Let A be a finite set (for simplicity) say $A = \{\alpha_1 < \alpha_2 < \alpha_3 < \cdots < \alpha_m\}$ and let $\forall i$, $D\alpha_i \cap D\alpha_{i+1} \neq \phi$ and $f_{\alpha i}\ |D\alpha_i \cap D\alpha_{i+1} = f\alpha_{i+1}|D\alpha_i \cap D\alpha_{i+1}$ (where $i = 1$ to $m - 1$) and the same can be shown as in the given figure.

So, here, $D_{\alpha 1}$ and $D_{\alpha 2}$ have a non-trivial intersection and $D_{\alpha 3}$ may or may not intersect with $D_{\alpha 2}$, but it has a non-trivial intersection with $D\alpha_2$ and so on till the non-trivial intersection between $D\alpha_{m-1}$ and $D\alpha_m$. Here, the function $f\alpha_1$ is analytic in $D_{\alpha 1}$ and the function $f_{\alpha 2}$ is analytic in $D_{\alpha 2}$ and so on with the values in $\mathbb{C}$. So, in other words, there is a chain of direct analytic continuation ($D_{\alpha 1}$, $f_{\alpha 1}$),...,($D_{\alpha m}$), $f_{\alpha m}$). Therefore, it is to be said that the ($D_{\alpha m}$, $f_{\alpha m}$) is the indirect analytic continuation of the first pair ($D_{\alpha 1}$, $f_{\alpha 1}$). Usually, the word ***indirect*** is omitted from general standard literature, but the reason for mentioning it is that if it is started with pair and has a chain such that the ending has a pair and has the same domain as the starting pair, the function obtained will be completely different. In this case, the function

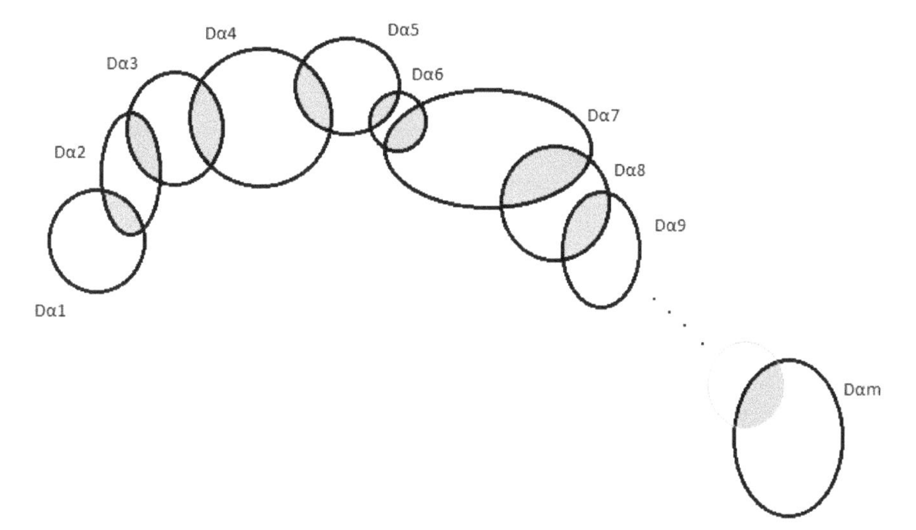

Figure 10.2 Indirect analytic continuation.

elements belong to the first pair and the last pair is obtained as a function of the first pair; these two are two different branches of analytic function. Hence, we arrive from one branch to another branch. This is the motivation and importance of the indirect analytic continuation concept.

Thus, on taking a general analytic function, if its derivative does not vanish at a point, then there is a neighborhood of that point. This function also has an inverse which is also analytic, and this is what the inverse function theorem states. This analytic function is one to one and has an inverse if its derivative does not vanish, but if the derivative vanishes, then the point is a critical point. But even at the neighborhood of the critical point, the branches of the inverse function can be obtained. They are functional inverses that are not actually inverse because in the neighborhood of the critical point, it will be a many-to-one function.

The process continues from one branch to another branch locally that comes via indirect analytic continuation. It is the importance of indirect analytic continuation that an indirect analytic continuation helps to find all the possible branches of a function (and, of course, here, it is to be noted that the branches of a function do have a singular point and it has a branch cut).

Hence, if an analytic continuation is not direct, then it is an indirect analytic continuation.

10.2.3 Complete Analytic Functions

An analytic function is considered to be a complete analytic function when it contains its each and every analytic continuation of function elements.

Let us take a pair (f, D), a function element. Here, f is a function which is analytic on domain D. Let us define a relation $\mathbb{R}$ on the set of all the function elements when f_2 is the analytic continuation of f_1 from domain D_1 to D_2 as

$$(f_1, D_1)\mathbb{R}(f_2, D_2).$$

Since it is known that the indirect analytic continuation is allowed, the relation $\mathbb{R}$ is said to be an equivalence relation and an equivalence class under the relation $\mathbb{R}$ of function elements is said to be the complete analytic function.

Let us consider some pair of function elements of the complete analytic function F, as (f_1, D_1) and (f_2, D_2) such that $f_1(z) \neq f_2(z)$ for some $z \in D_1 \cup D_2$. If there is an existence of the (f_1, D_1) and (f_1, D_2) function elements, then F is a multiform function; otherwise, F is not a multiform function.

A multiform function is a sort of more conventional and formal version of the known multi-valued function although it has an advantage that it breaks down into pieces in which it will be a single valued function. This multiformity makes an appearance because of the way these pieces fit in conjunction. A geometric approach to these proceeds toward the idea of Riemann surfaces.

10.2.4 Analytic Continuation Along Arcs

Analytic continuation along arcs is the continuation, which is along the curves such that it starts from an analytic function which is defined around a point and starts extending that function along an arc (curve) via analytic functions while those analytic functions can be defined on the small overlapping regions which cover those curves (as shown in Figure 10.3).

One of the most important results in this frame is ***Monodromy theorem.*** This theorem gives the idea to extend an analytic function all along a curve which starts from the original given domain of the function and ended up in the larger region (compared to the original one).

Here arises a potential problem that analytic continuation can be done by many ways (i.e., via many curves) which ended up in the larger region on the same point. The *Monodromy theorem* provides the condition, which is sufficient for analytic continuation, despite the curve which is used to reach the desired larger region at the point given for the same value; therefore, the resulting extension of the analytic function will be well-defined and not multi-valued.

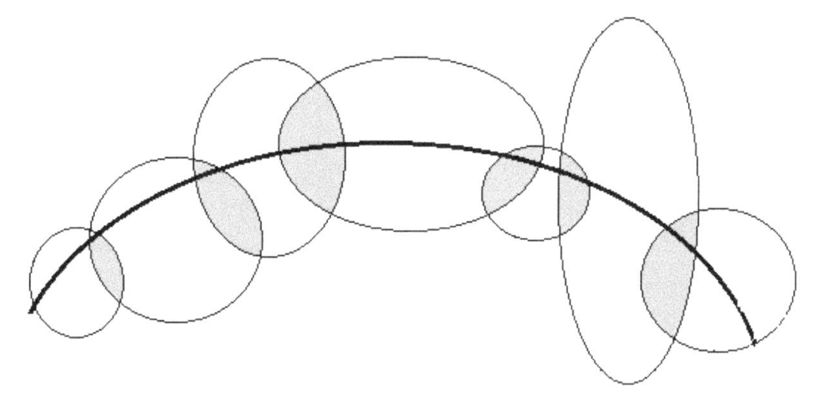

Figure 10.3 Analytic continuation along an arc.

10.3 Singularities

Let us define singularity in general, if any analytic continuation of a function f is not possible at a point z_o, then $z_o$will be considered as the *singularity* of the given analytic function. It is known that there exist various kinds of singularities such as poles, essential singularities, isolated singularities, and natural boundaries. But according to the new definitions, the singularities which are removable are not further considered as singularities because analytic continuation "filled" that missing point (singularity).

For the functions like *log* z and $\sqrt{z}$, we experienced a new kind of singularity, known as *branch point* (branch cuts).

Definition 10.2. *(Branch cuts). In the complex plane, a curve is considered as a branch cut such that it is feasible to explicate a single analytic branch of a multiform function on the complex plane where the curve is excluded.*

Brach cuts are informally said as branch points; these are the points where several sheets of multi-valued functions come together. Those various sheets of the given functions are said to be the *branches.*

Let us take a simple example,

$$f(z) = z^{1/2}.$$

This function f(z) has two branches; it has a square root one with a plus sign and another one with a minus sign.

Most of the times, branch cuts are taken between the pairs of the branch points but not always.

10.4 Riemann Surfaces

It is known that Weierstrass work was completely focused on power series and there is a lack of geometric viewpoint (as considering the published work). This deficiency was aided by wide-ranging ideas acquainted by *Bernhard Riemann* (1826–1866). Particularly, the idea of ***Riemann surfaces*** was introduced from 1851. In this concept, the multi-valued functions have been treated by splitting the complex plane into multiple layers, and on each layer, the function is single valued. An important feature is that the layers join up topologically and it is a crucial study as to how they are doing so.

10.4.1 The Idea of a Riemann Surface

The aim of the following discussion is to know about the generation of Riemann surface. For this, let us consider a real surface (like sphere, cylinder, or torus in R^3) and perform some complex analysis on it.

For example, we have the following:

1. Cylinder, $\mathbb{S}^1 \times \mathbb{R}$, where S^1 is the unit circle and R is the copy of the real line.
2. Two sphere, $\mathbb{S}^2$ which means there are two spheres in three spaces.
3. Torus, $\mathbb{T}^2$ which is same as $\mathbb{S}^1 \times \mathbb{S}^1$. It means that T^2 is homeomorphic to $S^1 \times \mathbb{S}^1$.

So, these are all the surfaces which can be imagined by sitting inside a surface of R^3, three-dimensional real space. Consider a point on such a surface (such as in Figure 10.4(a)) and take a small disc-like neighborhood which would be homeomorphic to the disc in the plane. To get a disc on a plane, cut out a disc-like neighborhood drawn around the considered point and flatten it (see Figure 10.4(b)).

As it can be seen, the surface is curved, but on taking a point on the surface and taking a small neighborhood about that point and flattens, it looks like a disc or a neighborhood of a point on the plane.

Now let us perform complex analysis on the given surface; doing complex analysis means defining the notion of what a holomorphic function is and what an analytic function is and then studying the properties of analytic functions. And, more importantly, let us explore how the analytic or holomorphic functions change as the objects changes.

In simple words, performing complex analysis helps to define and study the holomorphic functions on each of these surfaces, and let us analyze that

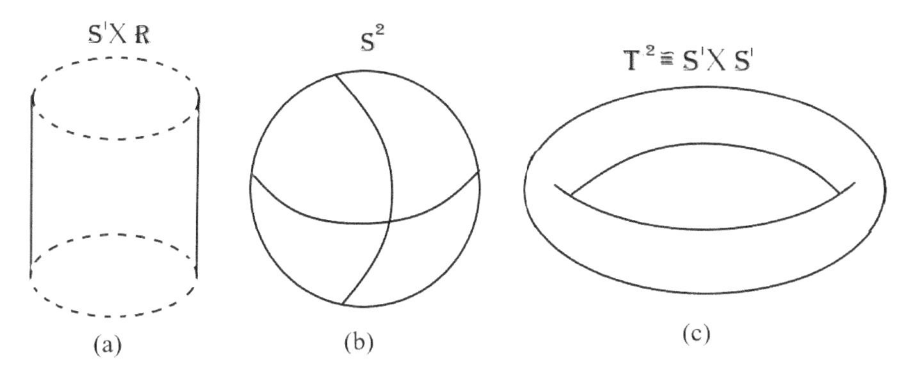

Figure 10.4 (a) Real surfaces: infinite cylinder (first), two spheres (second), and torus (third).

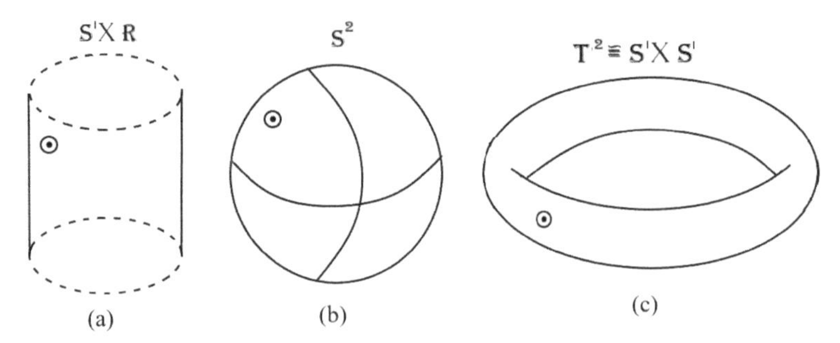

Figure 10.4 (b) Point taken on the real surfaces: infinite cylinder (first), two spheres (second), and torus (third).

such functions give us some information about the geometry of these objects and thus give us difference in these objects.

Some topological properties (see Figure 10.4(b)):

- It can be evident from the defined properties that in Figure 10.4(b), figure (b) and (c) are compact.

The figures (b) and (c) are special when compared to figure (a), in the sense that (b) and (c) are compact because both (b) and (c) are closed and bounded subset of $\mathbb{R}^3$. Here, (a) is an infinite cylinder; it is not bounded, and it is known that the subset of Euclidean spaces is compact if and only if it is closed and bounded. So, (a) is certainly closed, but it is not bounded, and, therefore, (a) is not compact.

- Also, it can be observed that the figure (b) is simply connected, but (a) and (c) are not simply connected.

Among the three figures, (b) is the surface that is simply connected; the notion of simply connected surface is the property that can be visualized by taking a loop on a surface and continuously shrinking that loop to a point without going away from the surface. Although, taking a loop on a sphere and continuously shrinking that loop to a point is possible on the sphere but it cannot be done always on a torus because if we tried to take a loop around the surface of the torus, it became a tubelike structure that cannot be shrinked to a point. Similarly, if a loop is taken on the outside surface of the torus, it can never be shrinked to a point because of the hole in between; so torus is not simply connected and, similarly, the infinite cylinder (a) is not simply connected.

Hence, it can be seen that there are many topological properties and going on to differential geometry is also possible where the curvature of the given object can be studied; so there are lot of geometric things that can be

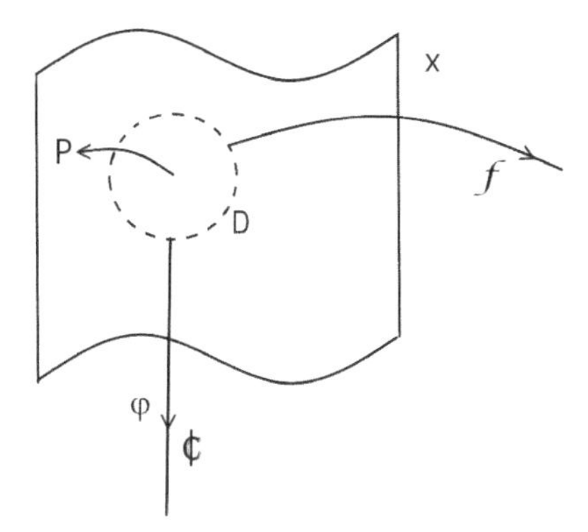

Figure 10.4 (c) Topological isomorphism.

done on these objects. Also, how properties of these functions can reflect these geometric properties is to be analyzed, and this is the purpose of doing analysis with geometry. Here, geometry means the topological properties of the object along with much more complicated structures.

Let X be the surface, and it can be cylinder, sphere, or the torus and there is a point *p* on the surface, D is the disc-like neighborhood (meaning it is homeomorphic to a disc on the complex plane). In formal mathematics, it can be said as chosen a homeomorphism or a topological isomorphism of this neighborhood of the unit disk in the complex plane where point *p* is the center of that disc.

Let the homeomorphism φ be defined on D into the complex plane C. So, φ (D) is a disc in the complex plane and φ: D $\rightarrow \varphi(D) \subset \mathbb{C}$ is a homeomorphism and φ (p) is the center of φ (D). Now, suppose there exists a complex valued function *f* defined on the disc D and the aim is to find when this complex valued function is analytic at the point p. To define the analytic function on the surface, the first step is to define analytic function at a point; once it is done, then the analytic function can be defined on an open set as well (see Figure 10.4(d)).

There is an image of the disk D and it goes to the point *φ(p)* that is the center of the disc φ *(D)*; this is how φ maps to p. It is known that φ is a homeomorphism or it is a topological isomorphism which leads to the conclusion that φ is bijective and continuous and the inverse of φ is also continuous. So, there is a continuous map in the direction of φ^{-1}, which is also a topological isomorphism. Therefore, going by φ^{-1} and then applying f, a composition f $\circ\varphi^{-1}$ was found. For such a function, it will be easy to define analyticity.

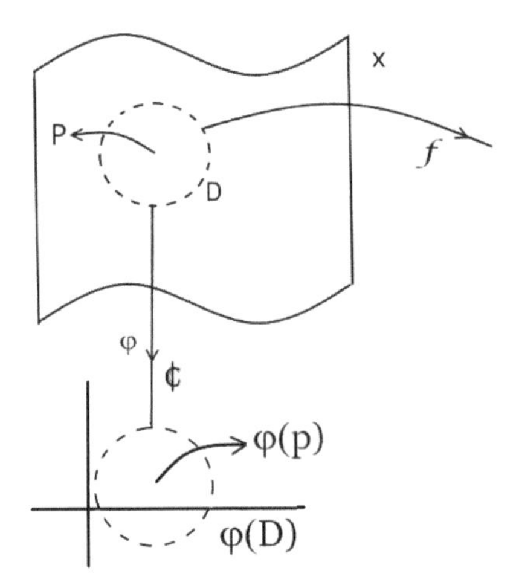

Figure 10.4 (d) Topological isomorphism $\varphi : D \to \varphi(D) \subset C$.

Formally, we define a function f: $D \to C$ given on a disc-like neighborhood D to be analytic at point $p \in D$ if, after choosing a homeomorphism φ: $D \to \varphi(D) \subset C$, the resulting function $\varphi(D) \to C$ followed by $f \circ \varphi^{-1}$ (means first go by φ^{-1} and then f) is analytic at φ(p).

The only ambiguity in this definition is that the φ could vary and we want to consider only those φ for which $f \circ \varphi^{-1}$ is analytic or holomorphic at φ(p) always.

Definition 10.3 *(Complex chart). A pair (U, φ) where $U \subset X$ is open and φ: $U \to \varphi(U) \subset \mathbb{C}$ is a homeomorphism onto an open subset φ (U) of $\mathbb{C}$ is called a complex coordinate chart for each point of U.*

For example (see Figure 10.4(d)), take the disc-like neighborhood surrounding the point p and take the homeomorphism of the disc-like neighborhood with a disc in the complex plane; then the pair (D, φ) is called the *coordinate chart* for every point of D. Because for every point of D, it gives you a point in the complex plane which has coordinates and it does it in a manner which is topological isomorphism (see Figure 10.4(e)).

(see Figure 10.4(e)) Let there exist a surface X and a point on the surface be x. Let U be the open set of X and (U_1, φ_1) be the coordinate chart into the complex plane, where φ_1 is the homeomorphism, and it will map the open set U_1 into some open set of the complex plane that is $\varphi_1(U_1)$. So, U_1 is a topological isomorphism of an open subset containing a point x with an open subset of complex plane (Figure 10.4(d) is the special case where D looks

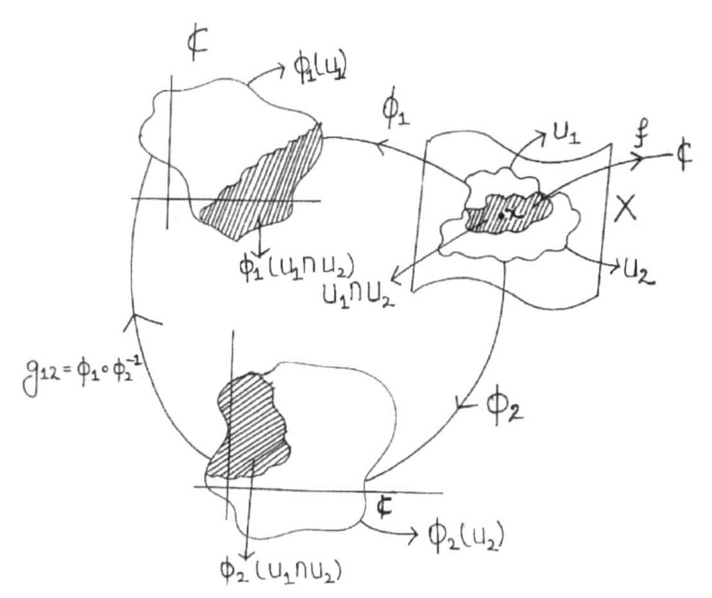

Figure 10.4 (e) φ_1 maps to $\varphi_1(U_1)$, φ_2 maps to $\varphi_2(U_2)$, and g_{12} is transition function.

like a disc and the image φ (p) was actually a disc). The most general set on which analytic function can be defined is an open set.

Suppose there exists another chart (U_2, φ_2), where U_2 is another open subset of X and φ_2 is another homeomorphism from U_2 into a subset of a complex plane. So, U_2 is another neighborhood of the point x and φ_2 is identification of that open neighborhood U_2 with an open subset φ_2 of U_2 in the complex plane.

Here, the neighborhoods of that point common to these two sets have a pair of coordinates because if $U_1 \cap U_2$ is considered, then for every point in $U_1 \cap U_2$ including x, if the image is taken under φ_1, it will give a coordinate, and if the image is taken under φ_2, then it will give another point, and, thus, there will be two coordinates. Take the shaded region, $U_1 \cap U_2$ (keep in mind that intersection of two open sets is again an open set), under homeomorphism φ_1, it will go to the shaded region, $\varphi_1(U_1 \cap U_2)$. In the same way, take the image of x under φ_2, it will get another open set, that is, φ_2 ($U_1 \cap U_2$). Therefore, a map can be defined as $\varphi_1(U_1 \cap U_2) \rightarrow \varphi_2(U_1 \cap U_2)$. And in the another map, firstly $\varphi_{2^{-1}}$ will be applied and then apply φ_1.

Now define g_{12},

$$g_{12} = \varphi_1 \circ \varphi_{2^{-1}}$$

(which means, firstly $\varphi_{2^{-1}}$ is applied and then φ_1 is applied).

So, g_{12} is φ_{2-1} restricted to $\varphi_2(U_1 \cap U_2)$ followed by φ_1 restricted to $\varphi_1(U_1 \cap U_2)$. Mathematically,

$$g_{12} = \varphi_1|(U_1 \cap U_2) \circ \varphi_{2-1}|\varphi_2(U_1 \cap U_2)$$

where the vertical line denotes restriction of a map to a subset. And g_{12} is said to be a "***Transition function***".

To know the importance of this transition function, assume that there is a function defined in a neighborhood of the point; to say that the function is analytic at that point, it is necessary to use a chart according to the definition. So, if there is a complex valued function f defined on $(U_1 \cap U_2)$, then the function is analytic at a point x, and among the two charts, any of them can be used.

Definition 10.4. *Let f be analytic at x w.r.t. the chart* (U_1, φ_1)*; if* $(f \circ \varphi_1{}^{-1})$ *is analytic at* $\varphi_1(x)$*, f is also analytic at x w.r.t. the chart* (U_2, φ_2) *if* $(f \circ \varphi_1{}^{-1})$ *is analytic at* $\varphi_2(x)$*.*

It follows that both are equivalent if g_{12} is holomorphic isomorphic, which happens if g_{12} is holomorphic. This is because

$$(f \circ \varphi_1{}^{-1}) \circ g_{12} = (f \circ \varphi_2{}^{-1})$$

$$(f \circ \varphi_1{}^{-1}) = (f \circ \varphi_2{}^{-1}) \circ g_{12}{}^{-1}$$

$$g_{12}^{-1} = g_{21}.$$

In conclusion of the above discussion, by doing analysis at a given surface, it gives a collection of charts with any point, and whenever two charts are intersecting, they must be compatible (i.e., the corresponding transition functions are holomorphic).

If X is given a collection of charts which cover every point of X and which are pairwise compatible (i.e., whenever two charts intersect, the transition function are holomorphic), then we may define and study holomorphic (analytic) function on X without ambiguity; such a collection of compatible charts on X is called a complex ***Atlas*** *on X and X together with a complex Atlas is called a* ***Riemann Surface.***

Riemann surface is the correct structure that allows doing complex analysis. There can be different Atlases giving different Riemann surfaces and structures on the same set X.

Examples:

A considerably basic example for Riemann surface is the complex plane C. Analytic continuation provides some non-compact Riemann surfaces; some of them are $f(z) = z$, $f(z) = z^{1/2}$, $f(z) = z^{1/3}$, $f(z) = z^{1/4}$, $f(z) = \arcsin z$, $f(z) = \log z$.

The Riemann surfaces provide a "global" view of functions instead of separating them into "local pieces".

10.5 Conclusion

In this chapter, we have discussed the role of analytic continuation in the complex analysis and how it can be used in the context of Riemann surfaces as well. In this, we have discussed what analytic continuation is and what its types are. Hence, it was found that such analytic continuations are not always unique, and the extension of the domains will not always be possible. By the indirect analytic continuation, we have learnt another way to extend the domains and it is useful whenever direct analytic continuation is not possible. *Monodromy theorem* is one of the most important results of analytic continuation along the arcs. If a point does not support the analytic continuation of a function, such points will be considered as poles, essential singularities, isolated singularities, and natural boundaries. *Hadamard gap theorem* is an important result with respect to the analytic continuation of the complex power series. The non-zero terms in that complex power series are in such an order that there will be a suitable gap between them. Such a power series cannot be extended as an analytic function on the boundary of its disc of convergence. *Gamma function* and *Riemann zeta function* are some of the major examples of analytic continuation. As discussed, the idea of Riemann surfaces was developed while exploring the maximal analytic continuation, but it was a geometrical approach. Talking about the geometric ideas, it is known that there are many practical applications of complex analysis and it is widely used by engineers and physicists in different contexts. For example, in the field of physics, it can be used for the *analytic continuation of potential fields*; that is, this concept was proved as especially useful tool in the field of *two-dimensional potential theory*. Analytic continuations have its connections to the *fluid mechanics* and also have application in the construction of *potential energy curves*. The method of analytic continuation was also used to study the *low-energy negative-ion states of beryllium* (configuration *2s2εpP2*) and *magnesium* (configuration *3s2εpP2*) atoms, i.e., to investigate the shape resonances of Be^- and Mg^-. The above discussed are a few examples and applications which show that the concept of analytic continuation is getting used in various research areas.

Acknowledgement

I would like to thank River Publications for giving me this great opportunity. They have supported me throughout patiently in order to complete this chapter.

References

[1] Ahlfors, L. V. (1979). Complex Analysis: An Introduction To The Theory Of Analytic Functions Of One Complex Variable. McGraw-Hill, Inc.
[2] Bieberbach, L. (1955). Analytische Fortsetzung. Springer-verlag.
[3] Chiang C. Mei. (1997). Mathematical Analysis In Engineering How To Use The Basic Tools. Cambridge University Press.
[4] Cohen, H. (2007). Complex Analysis With Applications In Science And Engineering (Second Ed.). Springer.
[5] Jones, W. B., & Thron, W. J. (1980). Continued Fractions: Analytic Theory And Application. Encyclopedia Of Mathematics And Its Applications 11.
[6] M. Kline. (1972). E. Mathematical Thought From Ancient To Modern Times. Oxford University Press.
[7] Narasimhan, R. (1973). Analysis On Real And Complex Manifolds (Second Ed.). North-holland Elsevier.
[8] Roel Snieder, and Kasper Van W I Jk. (2001). A Guided Tour Of Mathematical Methods For The Physical Sciences (First Ed.). Cambridge University Press.
[9] R. Remmert. (1998). Theory Of Complex Functions. Springer, New York.
[10] Sondow, Jonathan, & Weisstein, Eric W. (N.D.). "Riemann Zeta Function. Math World" A Wolfram Web Resource. Https://Mathworld.Wolfram.Com/Riemannzetafunction.Html
[11] Stefan Kranich. (2014, March 12). Computational Analytic Continuation. Vol. 1.
[12] Tavora, M. (N.D.). A Gentle Introduction To Analytic Continuation: How To Extend The Domain Of Analytic Functions. Medium.

[13] Theodore Gamelin. (2003). Complex Analysis. Springer.
[14] K. D. Jordan. (1975). Applications Of Analytic Continuation In The Construction Of Potential Energy Curves. International Journal Of Quantum Chemistry Symp. No. 9. 325–336 (1975).
[15] Roman Čurík, I. Paidarová, and J. Horáček. (2018). Shape Resonances Of Be- And Mg- Investigated With The Method Of Analytic Continuation. Physical Review A Covering Atomic, Molecular, And Optical Physics And Quantum Information.

11

Crime Prediction via Fuzzy Multi-Criteria Decision-Making Approach Under Hesitant Fuzzy Environment

Soumendra Goala, Palash Dutta, and Bulendra Limboo

Department of Mathematics, Dibrugarh University,
Dibrugarh 786004, India
E-mail: soumendragoala@gmail.com; palash.dtt@gmail.com;
rs_bulendralimboo@dibru.ac.in

Abstract

It can be observed that since a few years, crime in some cities has been increase data higher rate, which causes a major problem and leaves a very bad impact on society. Too much engagement with criminal cases is obviously a time- and energy-consuming process that slows down the progress of a society. For this reason, predicting crimes through efficient methodologies and artificial intelligence is becoming hot as well as a necessary issue for a few years. In this chapter, a methodology has been introduced to predict the next crime under a hesitant fuzzy environment. Also, a numerical case study has been carried out to show the applicability of the methodology.

Keywords: Hesitant fuzzy set, Distance measure, serial crime, crime prediction, Fuzzy set, Hesitant fuzzy element, Resemblance measure, Score function, Deviation function.

11.1 Introduction

For ages, the process of criminal investigation and crime prediction is always a decision-making situation under uncertainty due to various factors such as the absence of evidence, presence of several conflicting pieces of evidence, a different perspective of different investigators, presence of different eyewitnesses, and many more. In addition, the task of expressing crimes mathematically is difficult due to the linguistic expressions used in the investigation reports by different investigators and different interpretations of the same type of evidence in the crimes. Also, the increase in crime results in a huge workload for law enforcement agencies. As a consequence, all these factors obstruct the law enforcement process of any country or system, which is very bad for the society. For this reason, modeling of criminal investigation and prediction being tricky, it is always necessary to have efficient methodologies, modeling, or decision support systems for predicting crimes. For these reasons, a study has been carried out in the field of fuzzy multi-criteria decision-making methodology to address the issue of uncertainty in crime prediction more efficiently. In this chapter, hesitant fuzzy sets have been utilized to express the crimes and possible crime spots and a methodology has been introduced to predict the next crime under a hesitant fuzzy environment by utilizing the distance measure on hesitant fuzzy sets. Also, a numerical case study has been carried out to show the applicability of the methodology.

From the evolvement of fuzzy set theory [7], researchers have begun to incorporate fuzzy set theory in the decision-making process [11] due to its enormous applicability to deal with an uncertain environment. In literature, there are only a few studies that have been found solely devoted to crime prediction and serial crime detection by using the fuzzy set theory. An intuitionistic fuzzy multi-criteria decision-making approach based on the intuitionistic fuzzy sets (IFSs) has been discussed to identify the serial crimes from a large volume of crimes by Goala and Dutta [12]. They introduced a new concept of resemblance measure for IFSs and used resemblance measure for identifying the serial crimes. Goala and Dutta [13] studied the detection of serial crimes via a novel distance measure under the hesitant fuzzy environment in a pair-wise comparison process. A fuzzy clustering technique was developed by Adeyiga and Bello [4] for criminal profiling to detect and stop crimes. Gupta and Kumar [14] used some properties of crimes in a city and mapped by the area using fuzzification and later the value that is evaluated by defuzzification to detect the serial crimes. Sheng *et al.* [16] introduced a method to detect the crime patterns for efficient routine for police duty deployment. A new idea of using fuzzy time series for the prediction of crimes was developed by Shrivastav and Ekta [1]. A Multi Attribute Decision Making (MADM) approach was developed by Albertetti *et al.* [2] to analyze the serial crime of high volume crimes. Grubesic [15] put forward a method to detect the crime-prone spot in a city through the fuzzy clustering analysis.

11.2 Preliminaries

Fuzzy Set [7]: A fuzzy set A on universal set X is defined as the collection of pairs

$$A = \{(x, \mu_A(x)) : x \in X\},$$

where $\mu_A : X \to [0,1]$ is the membership function on X or grade of membership value.

Hesitant Fuzzy Sets [17]: A hesitant fuzzy set on X is defined as the function which is a subset of $[0,1]$ when applied to X. More precisely, an expression of Hesitant Fuzzy Sets (HFSs) A on X is given by

$$A = \{\langle x, h_A(x)\rangle : x \in X\}, \quad \text{where } h_A(x) \subseteq [0,1].$$

Here, the membership value $h_A(x)$ is called the hesitant fuzzy element (HFE in short) and it represents all the possible and distinct membership grade of $x \in X$ in the set A.

Score Function [8]: The score function $s(h)$ of an HFE h is defined as

$$s(h) = \frac{1}{l_h} \sum_{\gamma \in h} \gamma,$$

where l_h is the number of elements in h.

Based on the concept of score functions of HFEs, two HFEs can be compared. Suppose, for any two HFEs h_1 and h_2 with their score functions $s(h_1)$ and $s(h_2)$, we have

a) $s(h_1) > s(h_2) \Rightarrow h_1 > h_2$;
b) $s(h_1) = s(h_2) \Rightarrow h_1 = h_2$.

From the definition, it is noted that the higher the average results, the greater the score value and accordingly higher the HFE. However, this definition of score function fails to compare between two HFEs since two different HFEs may have the same score value. To overcome the drawback, Chen and Xua [9] introduced the idea of deviation degree and utilized it to compare between two HFEs where score functions fail.

Deviation Function [4]: The deviation degree of an HFE h is defined as

$$\overline{\sigma}(h) = \left(\frac{1}{l_h} \sqrt{\sum_{\gamma \in h} (\gamma - s(h))^2} \right)^{\frac{1}{2}}.$$

Chen and Xua [9] introduced a new technique to compare the HFEs by utilizing the deviation function.

Consider h_1 and h_2 be two HFEs with the score functions $s(h_1)$, $s(h_2)$, and $\overline{\sigma}(h_1)$, $\overline{\sigma}(h_2)$ be its deviation degree, respectively. Then, the following properties hold:

a) $s(h_1) > s(h_2) \Rightarrow h_1 > h_2$
b) $s(h_1) = s(h_2)$

(1) $\overline{\sigma}(h_1) = \overline{\sigma}(h_2) \Rightarrow h_1 = h_2$
(2) $\overline{\sigma}(h_1) > \overline{\sigma}(h_2) \Rightarrow h_1 < h_2$
(3) $\overline{\sigma}(h_1) < \overline{\sigma}(h_2) \Rightarrow h_1 > h_2$

11.3 Distance Measures

Distance measure has been widely used to measure the degree of distance or difference between two information or arguments. For these types of usability, distance measure has been utilized in various types of multi-criteria decision-making. There are many distance measures proposed by many researchers and used in different algorithms for the decision-making approaches in the field of fuzzy set theory due to the reliability and flexibility to cope with the uncertainty that occurs in the course of decision-making.

A distance measure between two sets A and B can be understood as a function $d : A \times B \rightarrow [0, 1]$.

Let $A = \{\langle x_i, h_A(x_i)\rangle : x_i \in X\}$ and $B = \{\langle x_i, h_B(x_i)\rangle : x_i \in X\}$ be any two HFSs. The distance measures $d(A, B)$ will satisfy the properties given below:

(1) $0 \leq d(A, B) \leq 1$
(2) $d(A, B) = d(B, A)$
(3) $d(A, B) = 0 \Leftrightarrow A = B$

11.3.1 Some Well-Known Distance Measures on Hesitant Fuzzy Sets

Many researchers have studied various distance measures with the application in different fields. Some of them are given below.

Xu and Xia [19] proposed a generalized hesitant normalized distance and defined it as

$$d(A, B) = \frac{1}{n}\sum_{i=1}^{n} w_i \left[\sum_{j=1}^{l_{x_i}} \left|h_A^{\sigma_j}(x_i) - h_B^{\sigma_j}(x_i)\right|^{\lambda}\right]^{\frac{1}{\lambda}}, \quad \lambda > 0$$

where l_{x_i} is the cardinality of the ith HFEs, $h_A^{\sigma_j}(x_i)$, and $h_B^{\sigma_j}(x_i)$ are the jth largest value in the HFE in A and B, respectively, and the HFEs are arranged in the decreasing order.

In particular, for fixed $\lambda = 1$, the above distance $d(A, B)$ is referred to as the Hesitant weighted Hamming distance measure, and for $\lambda = 2$ is called $d(A, B)$ Hesitant weighted Euclidean distance measure.

Again, the generalized weighted Hausdorff distance measure is defined as

$$d(A, B) = \frac{1}{n}\sum_{i=1}^{n} w_i \left[\max_j \left|h_A^{\sigma_j}(x_i) - h_B^{\sigma_j}(x_i)\right|^{\lambda}\right]^{\frac{1}{\lambda}}, \quad \lambda > 0.$$

For the fixed $\lambda = 1$ and $\lambda = 2$, the distance $d(A, B)$ is successively defined as the weighted Hausdorff Hamming distance measure and weighted Hausdorff Euclidean distance measure, respectively.

Also, Xu andXia [19] advanced a generalized hybrid hesitant weighted distance measure joining the generalized hesitant weighted distance measure and the generalized hesitant weighted Hausdorff distance measure as

$$d(A, B) = \frac{1}{2n}\sum_{i=1}^{n} w_i \left[\sum_{j=1}^{l_{x_i}} \left|h_A^{\sigma_j}(x_i) - h_B^{\sigma_j}(x_i)\right|^{\lambda} + \max_j \left|h_A^{\sigma_j}(x_i) - h_B^{\sigma_j}(x_i)\right|^{\lambda}\right]^{\frac{1}{\lambda}}, \quad \lambda > 0.$$

For the fixed $\lambda = 1$, $d(A, B)$ is the hybrid weighted hesitant Hamming distance measure, and for the fixed $\lambda = 2$, $d(A, B)$ is the hybrid weighted hesitant Euclidean distance measure.

Now, it can be noticed that hesitant fuzzy sets do not have an equal number of elements in HFEs. To evaluate distance between two HFSs having unequal number membership grades decision-makers add extra elements in the deficit HFEs. To overcome the disadvantages of adding the extra elements in the deficit of HFEs, Goala and Dutta [13] introduced a new type of distance measures in their study as follows:

$$d(A, B) = \sum_{i=1}^{l} w_i \frac{|s(h_A(x_i)) - s(h_B(x_i))| + |\overline{\sigma}(h_A(x_i)) - \overline{\sigma}(h_B(x_i))|}{1 + |s(h_A(x_i)) - s(h_B(x_i))| + |\overline{\sigma}(h_A(x_i)) - \overline{\sigma}(h_B(x_i))|},$$

where $l = |A| = |B|$, $h_A(x_i)$, and $h_B(x_i)$ are the HFEs of x_i, $s(h_A(x_i))$ and $s(h_B(x_i))$ are the score functions, and $\overline{\sigma}(h_A(x_i))$ and $\overline{\sigma}(h_B(x_i))$ are the variation functions of the HFEs $h_A(x_i)$ and $h_B(x_i)$, respectively.

11.4 Multi-Criteria Decision-Making Approach for Crime Prediction

Consider a set C of alternatives consisting of n crimes such that $C = \{C_1, C_2, \ldots, C_n\}$ and let A be the set of m characteristics such that $A = \{A_1, A_2, \ldots, A_m\}$ of crime scenes including action, behavior, and situational elements with weights $w = (w_1, w_2, \ldots, w_m)^T$, respectively. Therefore, a fuzzy multi-criteria decision-making situation with m criteria A_i (characteristics) and n alternatives, namely C_j, is expressed by the decision matrix put forward in [3]:

$$\begin{array}{c} \\ A_1 \\ A_2 \\ \vdots \\ A_m \end{array} \begin{array}{c} \begin{array}{cccc} C_1 & C2 & \ldots & C_n \end{array} \\ \begin{bmatrix} h_{11} & h_{12} & \ldots & h_{1n} \\ h_{21} & h_{22} & \ldots & h_{2n} \\ \vdots & \vdots & \ddots & \vdots \\ h_{m1} & h_{m2} & \ldots & h_{mn} \end{bmatrix} \end{array}.$$

Here, h_{ij} are the HFEs which give the degree to which certain characteristics A_i related by the criminal case C_j. The criminal case C_j is represented by hesitant fuzzy sets as follows:

$$C_j = \{\langle A_i;\, h_{ij}\rangle : A_i \in A\}, \text{ where } h_{ij} \to [0,1];\ i = 1, 2, \cdots, m \text{ and } j = 1, 2, \cdots, n.$$

Similarly, the next possible situation for crimes say $C_p = \{C_{p1}, C_{p2}, \cdots, C_{ps}\}$ can be expressed in terms of HFSs as

$$C_{pj} = \{\langle A_i;\, h_{pij}\rangle : A_i \in A\}, \text{ where } h_{ij} \to [0,1];\ i = 1, 2, \cdots, m \text{ and } j = 1, 2, \cdots, s.$$

Then, the satisfaction degree of the situation C_{pj} to be the next possible crime is

$$S(C_{pj}) = \frac{\left| \frac{1}{s}\sum_{j=1}^{s}\left\{\frac{1}{n}\sum_{i=1}^{n} d(C_i, C_{pj})\right\} - \frac{1}{n}\sum_{i=1}^{n} d(C_i, C_{pj}) \right|}{\frac{1}{s}\sum_{j=1}^{s}\left\{\frac{1}{n}\sum_{i=1}^{n} d(C_i, C_{pj})\right\}},$$

where d is the distance between two hesitant fuzzy sets.

The satisfaction degree of the next possible crime spot will signify to what extent the next crime spot may be the actual crime spot.

11.5 A Hypothetical Case Study

In this section, a methodology has been carried out and validated in the case study of credit card fraud detection. For this study, a portion of the dataset is taken which was originally studied by Goala and Dutta [12].

In this study, the following main criteria have been taken, which are generally used by criminals:

(1) A_1: Tricky phone calls or emails and asking for information about credit or debit cards.
(2) A_2: A virus or Trojan software sent through the email to access the information of the credit card holder or debit cardholder.
(3) A_3: The click jacking technique: The clicks are directed to some fraud website or advertisements or payment gateway or shopping website.
(4) A_4: Physical stealing of card information, e.g., card cloning while shopping.

Consider the three credit card frauds executed in a town and expressed as the hesitant fuzzy set as follows:

$$C_1 = \{\langle A_1, \{0.7, 0.7, 0.9\}\rangle, \langle A_2, \{0.7, 0.7, 0.9\}\rangle, \langle A_3, \{0.3, 0.3, 0.1\}\rangle, \langle A_4, \{0.7, 0.5, 0.9\}\rangle\}$$

$$C_2 = \{\langle A_1, \{0.7, 0.9, 0.7\}\rangle, \langle A_2, \{0.9, 0.9, 0.9\}\rangle, \langle A_3, \{0.5, 0.5, 0.3\}\rangle, \langle A_4, \{0.7, 0.5, 0.9\}\rangle\}$$

$$C_3 = \{\langle A_1, \{0.7, 0.7, 0.7\}\rangle, \langle A_2, \{0.7, 0.9, 0.9\}\rangle, \langle A_3, \{0.3, 0.1, 0.1\}\rangle, \langle A_4, \{0.5, 0.9\}\rangle\}$$

Similarly, consider seven people in the same town having habits or history of the following:

a) answering fake and unnecessary call regarding money;
b) history of virus attack on his computer;
d) clicking random advertisements;
e) shopping usually using credit cards.

The information of seven people and their degree of habits and history, and, as a consequence, the possibility of being targeted by earlier criminals depending on the above criteria has been given by the following hesitant fuzzy sets:

$$S_1 = \{\langle A_1, \{0.1, 0.3\}\rangle, \langle A_2, \{0.5, 0.3, 0.3\}\rangle, \langle A_3, \{0.7, 0.7, 0.9\}\rangle, \langle A_4, \{0.3, 0.3\}\rangle\}$$

$$S_2 = \{\langle A_1, \{0.5, 0.5, 0.3\}\rangle, \langle A_2, \{0.5, 0.3, 0.3\}\rangle, \langle A_3, \{0.7, 0.7, 0.7\}\rangle, \langle A_4, \{0.3, 0.5, 0.3\}\rangle\}$$

$$S_3 = \{\langle A_1, \{0.3, 0.3\}\rangle, \langle A_2, \{0.5, 0.3, 0.1\}\rangle, \langle A_3, \{0.7, 0.9, 0.9\}\rangle, \langle A_4, \{0.3, 0.5, 0.3\}\rangle\}$$

$$S_4 = \{\langle A_1, \{0.3, 0.3\}\rangle, \langle A_2, \{0.3, 0.5, 0.3\}\rangle, \langle A_3, \{0.7, 0.7, 0.5\}\rangle, \langle A_4, \{0.7, 0.5, 0.7\}\rangle\}$$

$$S_5 = \{\langle A_1, \{0.5, 0.3\}\rangle, \langle A_2, \{0.5, 0.3, 0.3\}\rangle, \langle A_3, \{0.7, 0.7, 0.7\}\rangle, \langle A_4, \{0.3, 0.5, 0.3\}\rangle\}$$

$$S_6 = \{\langle A_1, \{0.1, 0.3, 0.3\}\rangle, \langle A_2, \{0.5, 0.3, 0.1\}\rangle, \langle A_3, \{0.9, 0.9, 0.7\}\rangle, \langle A_4, \{0.3\}\rangle\}$$

$$S_7 = \{\langle A_1, \{0.7, 0.9, 0.7\}\rangle, \langle A_2, \{0.9, 0.9, 0.9\}\rangle, \langle A_3, \{0.5, 0.3, 0.3\}\rangle, \langle A_4, \{0.9, 0.9, 0.5\}\rangle\}.$$

Now the satisfaction degree has been evaluated for each people S_j to know the possibility of being targeted by criminals depending upon the earlier crimes executed using the distance measure introduced by Goala and Dutta [5] as follows:

$$S(S_j) = \frac{\left| 1 + \frac{1}{7}\sum_{j=1}^{7}\left\{\frac{1}{n}\sum_{i=1}^{3} d(C_i, S_j)\right\} - \frac{1}{3}\sum_{i=1}^{3} d(C_i, S_j) \right|}{1 + \frac{1}{7}\sum_{j=1}^{7}\left\{\frac{1}{3}\sum_{i=1}^{3} d(C_i, S_j)\right\}}.$$

The possibility of being targeted by criminals who executed the crimes C_1, C_2, and C_3 in terms of the satisfaction degree of each person depending on their behavior and past actions is evaluated below:

$$S(S_1) = \frac{\left| 1 + \frac{1}{7}\sum_{j=1}^{7}\left\{\frac{1}{n}\sum_{i=1}^{3} d(C_i, S_j)\right\} - \frac{1}{3}\sum_{i=1}^{3} d(C_i, S_1) \right|}{1 + \frac{1}{7}\sum_{j=1}^{7}\left\{\frac{1}{3}\sum_{i=1}^{3} d(C_i, S_j)\right\}}$$

$$= \frac{|1 + 0.705522 - 0.650272|}{1 + 0.705522} = 0.553936$$

$$S(S_2) = \frac{\left| 1 + \frac{1}{7}\sum_{j=1}^{7}\left\{\frac{1}{n}\sum_{i=1}^{3} d(C_i, S_j)\right\} - \frac{1}{3}\sum_{i=1}^{3} d(C_i, S_2) \right|}{1 + \frac{1}{7}\sum_{j=1}^{7}\left\{\frac{1}{3}\sum_{i=1}^{3} d(C_i, S_j)\right\}}$$

$$= \frac{|1 + 0.705522 - 0.771399|}{1 + 0.705522} = 0.624957$$

$$S(S_3) = \frac{\left| 1 + \frac{1}{7}\sum_{j=1}^{7}\left\{\frac{1}{n}\sum_{i=1}^{3} d(C_i, S_j)\right\} - \frac{1}{3}\sum_{i=1}^{3} d(C_i, S_3) \right|}{1 + \frac{1}{7}\sum_{j=1}^{7}\left\{\frac{1}{3}\sum_{i=1}^{3} d(C_i, S_j)\right\}}$$

$$= \frac{|1 + 0.705522 - 0.655478|}{1 + 0.705522} = 0.556989$$

$$S(S_4) = \frac{\left| 1 + \frac{1}{7}\sum_{j=1}^{7}\left\{\frac{1}{n}\sum_{i=1}^{3} d(C_i, S_j)\right\} - \frac{1}{3}\sum_{i=1}^{3} d(C_i, S_4) \right|}{1 + \frac{1}{7}\sum_{j=1}^{7}\left\{\frac{1}{3}\sum_{i=1}^{3} d(C_i, S_j)\right\}}$$

$$= \frac{|1 + 0.705522 - 0.733894|}{1 + 0.705522} = 0.602966$$

$$S(S_5) = \frac{\left| 1 + \frac{1}{7}\sum_{j=1}^{7}\left\{\frac{1}{n}\sum_{i=1}^{3} d(C_i, S_j)\right\} - \frac{1}{3}\sum_{i=1}^{3} d(C_i, S_5) \right|}{1 + \frac{1}{7}\sum_{j=1}^{7}\left\{\frac{1}{3}\sum_{i=1}^{3} d(C_i, S_j)\right\}}$$

$$= \frac{|1 + 0.705522 - 0.663488|}{1 + 0.705522} = 0.561685$$

$$S(S_6) = \frac{\left| 1 + \frac{1}{7}\sum_{j=1}^{7}\left\{\frac{1}{n}\sum_{i=1}^{3} d(C_i, S_j)\right\} - \frac{1}{3}\sum_{i=1}^{3} d(C_i, S_6) \right|}{1 + \frac{1}{7}\sum_{j=1}^{7}\left\{\frac{1}{3}\sum_{i=1}^{3} d(C_i, S_j)\right\}}$$

$$= \frac{|1 + 0.705522 - 0.633433|}{1 + 0.705522} = 0.544063$$

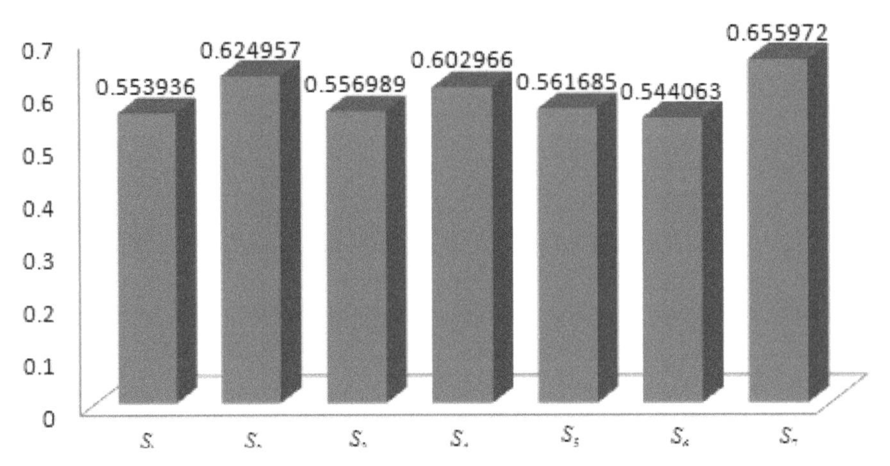

Figure 11.1 Satisfaction degrees of different persons of being next target.

$$S(S_7) = \frac{\left|1 + \frac{1}{7}\sum_{j=1}^{7}\left\{\frac{1}{n}\sum_{i=1}^{3} d(C_i, S_j)\right\} - \frac{1}{3}\sum_{i=1}^{3} d(C_i, S_7)\right|}{1 + \frac{1}{7}\sum_{j=1}^{7}\left\{\frac{1}{3}\sum_{i=1}^{3} d(C_i, S_j)\right\}}$$

$$= \frac{|1 + 0.705522 - 0.830689|}{1 + 0.705522} = 0.65972.$$

Thus, from Figure 11.1, it can be seen that the person S_7 has the highest satisfaction degree for the earlier frauds executed in the same town. Therefore, it can be concluded that the person S_7 may be the next target of the offenders who executed the crimes C_1, C_2, and C_3.

11.6 Conclusion and Discussion

Predicting a crime is becoming a major issue for intelligence agencies and law enforcement due to increased crimes in the last few years. It is always necessary to have efficient methodologies to detect serial crimes and predicting crimes to save time and energy for law enforcement agencies. For this reason, a study has been carried out to show a methodological and scientific approach for predicting serial crimes.

In this chapter, a methodology has been introduced to predict the next crime spot under a hesitant fuzzy environment via distance measure. In addition, a numerical case example has been demonstrated to describe the applicability of the methodology.

References

[1] A.K. Shrivastav & D. Ekta, Applicability of soft computing technique for crime forecasting: A preliminary investigation, International Journal of Computer Science & Engineering Technology, 3(9), pp. 415–421, 2012.

[2] F. Albertetti, P. Cotofrei, L. Grossrieder, O. Ribaux & K. Stoffel, The CriLiM Methodology: Crime linkage with a fuzzy MCDM approach, In European Intelligence and Security Informatics Conference, 2013.

[3] J. Klir & B. Yuan, Fuzzy sets and fuzzy Logic, Pearson Education, 2009.

[4] J.A. Adeyiga, & A. Bello, A review of different clustering techniques in criminal profiling, International Journal of Advanced Research in Computer Science and Software Engineering, 6(4), pp. 659–666, 2016.

[5] J.M. Sundaram, R. Karthikeyan & K. Raj, A survey of fuzzy based ARM clustering on crime pattern discovery, International Journal of Scientific & Engineering Research, 5(5), pp. 526–530, 2014.

[6] K. Stofel, P. Cotofrei & D. Han, Fuzzy clustering based methodology for multidimensional data analysis in computational forensic domain, International Journal of Computer Information Systems and Industrial Management Applications, 4, pp. 400–410, 2012.

[7] L. AZadeh, Fuzzy sets, Information and Control, 8(3), pp. 338–353, 1965.

[8] M.Z. Xia, Z. Xu & B. Zhu, Geometric Bonferroni means with their application in multi-criteria decision-making, Knowledge-Based Systems, 40, pp. 88–100, 2013.

[9] N. Chen, & X. Xua, Correlation coefficients of hesitant fuzzy sets and their applications to clustering analysis, Applied Mathematical Modelling, 37(4), pp. 2197–2211, 2013.

[10] N.H. Shamsuddin, S. Othman & H. Selamat, Identification of potential crime area using analytical hierarchy process (AHP) and geographical Information System (GIS), International Journal of Innovative Computing, 01(1), pp. 15–22, 2012.

[11] R.E. Bellman & L.A. Zadeh, Decision-making in a fuzzy environment, Management Science, 17(4), pp. 141–164, 1970.
[12] S. Goala & P. Dutta, Intuitionistic fuzzy multi-criteria decision-making approach to crimeCrime linkage using resemblance function, International Journal of Applied and Computational Mathematics, 5:112, 2019.
[13] S. Goala, P. Dutta, A fuzzy multi-criteria decision-making approach to crimeCrime linkage, International Journal of Information Technologies and Systems Approach, 11(2), pp. 31-50, 2018.
[14] S. Gupta, & S. Kumar, Crime detection and prevention using social network analysis, International Journal of Computers and Applications, 126(6), pp. 14–19, 2015.
[15] T.H. Grubesic, On the application of fuzzy clustering from crime hotspot detection, Journal of Quantitative Criminology, 22(1), pp. 77–105, 2006.
[16] T.L. Sheng, C.K. Shu & C.T. Fu, An intelligent decision support model using FSOM and rule extraction for crime prevention, Expert Systems with Applications, 37(10), pp. 7108–7119, 2010.
[17] V. Torra & Y. Narukawa, On hesitant fuzzy sets and decision, In the 18th IEEE International Conference on Fuzzy Systems, Jeju Island, pp. 1378-1382, 2009.
[18] V. Torra, Hesitant fuzzy sets, International Journal of Intelligent Systems, 25(6), pp. 529–539, 2010.
[19] Z.S. Xu & M.M. Xia, Distance and similarity measures for hesitant fuzzy sets, Information Sciences, 52(11), pp. 2128–2138, 2011.
[20] Z.S. Xu & M.M. Xia, Hesitant fuzzy information aggregation in decision-making, International Journal of Approximate Reasoning, 52(3), pp. 395–407, 2011.

12

Variable Selection in Bioinformatics: Methods and Algorithms

Mahak Bhushan[1], Shubham Kulkarni[2], Sonal Modak[3], and Jayaraman Valadi[4]

[1]Indian Institute Science Education and Research, Kolkata
[2]Centre for Modeling and Simulation, SPPU, Pune, India
[3]Lead Domain Engineer, Life Sciences and Healthcare Unit, Persistent Systems Inc., Santa Clara, CA 95054, USA
[4]Center for Informatics, School of Natural Sciences (SoNS), Shiv Nadar University, Greater Noida, Uttar Pradesh 201314, India
and
Computer Science Department, Flame University, Pune, India
E-mail: mahakbhushan22@gmail.com; kulkarnishubham01@gmail.com; samurai.modak@gmail.com; jayaraman.valadi@snu.edu.in; jayaraman.vk@flame.edu.in

Keywords: Feature selection, support vector machines (SVMs), kernel functions, hyperplane equation, random forest (RF), deterministic methods, heuristic methods.

12.1 Introduction

In Chemo & Bioinformatics problems, machine learning algorithms have been found to be very useful. In these problems, both supervised and unsupervised techniques are frequently employed. For example, in protein function annotation, it is possible to extract different types of features including sequence-based, structure-based, expression profile-based, composition-based, etc. In these large compendia of features, many may be noisy and may not contain information related to the function annotation problem at hand. With the presence of these irrelevant and noisy features, the classifier or regressor may not be able to build an optimal model and the performance may not be very good. This is because the model finds it difficult to differentiate between signal and noise. With the increase in the information content of the data set, the prediction accuracy will also increase and provide better generalization. Over fitting will also reduce with increase in informative features. The computational complexity will decrease, and the speed of training and testing will increase manifold. In bioinformatics, model interpretability increases with the usage of a small subset of relevant features and may provide important domain information in the form of identifiable biomarkers. Due to the importance of feature selection, several methodologies are available for feature selection. In this chapter, we discuss in detail different classes of feature selection methods and algorithms and their applications in Chemo & Bioinformatics. Feature selection techniques can be broadly classified as filter, wrapper, and embedding techniques.

12.2 Classification Methods

Several classification algorithms are routinely used for evaluating the performance of feature selection methods. These include support vector machines (SVM), decision trees, random forests (RF), logistic regression, Naïve Bayes classifier, neural networks, etc. Among these classifiers, SVM and RF have been found to be both robust and accurate. In this section, a brief introduction to SVM and RF is provided.

12.2.1 Support Vector Machine

SVM is a high-performance hyperplane-based classification algorithm [1]. The linear hyperplane maximizes the distance between itself and the nearest samples of both classes. This hyperplane with maximum margin can separate only linearly separable samples. To enable classification of linearly non-separable examples, the algorithm 1) takes the samples to a feature space with a higher dimension and 2) uses the linear hyperplane for separation. Employment of kernel functions increases computational simplicity and tractability. Computational costs are simultaneously reduced to a large degree. Further, SVM is a quadratic optimization formulation ascertaining a single global optimum solution. This attractive feature is exploited by several researchers in different groups for solving problems in various domains.

12.2.2 Random Forest Algorithm

RF is a forest or collection of unpruned decision tree algorithms. The algorithm was originally developed by Breiman [2]. It contains two different methods of randomness. In the first, bootstrap sampling with replacement is used in each tree, while in the second method, only a random subset of attributes is used for node splitting. Due to bootstrap sampling with replacement, roughly one-third of the examples are unused in each tree. These samples are known as out-of-bag examples. By virtue of this RF can use out-of-bag examples for testing algorithm performance. After model building, class prediction is done by majority voting. To obtain higher accuracy, optimal correlation among the trees needs to be maintained. This may be done by identifying and tuning the subset of features used in node splitting.

12.3 Performance Measures

Although accuracy is a traditional measure of performance, there are situations where it will not be appropriate. For certain problems, positive accuracy may be the desired performance measure, while for certain other problems, negative accuracy may be the desired one. Some problems may require maximization of the positive accuracy; on the other hand, some consider negative accuracy as an accurate measure of performance. For imbalance classification data sets with a highly skewed distribution of examples, class accuracy measures will fail to identify the correct model and model parameters. In these cases, we need to obtain a measure with the best tradeoff between both accuracies.

Examples with positive class labels predicted as positive are true positive examples. Examples with positive class labels predicted as negatives are false negatives. Examples with negative class labels predicted as negatives are true negative examples. Examples with negative class labels predicted as positive are false positive examples.

It is now possible to define various performance measures. True positive rate (TPR) or sensitivity is given as

$$\textit{True Positive Rate} = \frac{\textit{No. of True Positives}}{(\textit{No. of True Positives} + \textit{No. of False Negatives})}. \tag{12.1}$$

True negative rate or specificity is given as

$$\textit{True Negative Rate} = \frac{\textit{No. of True Negatives}}{\textit{(No. of False Positives + No. of True Negatives)}}. \quad (12.2)$$

The formula for positive predictive value or precision is given as

$$\textit{Positive Predictive Value} = \frac{\textit{No. of True Positives}}{\textit{(No. of True Positives + No. of False Negatives)}}. \quad (12.3)$$

The harmonic mean of precision and sensitivity makes up the F1 score and is given as

$$\textit{F1 Score} = \frac{2 * \textit{(No. of True Positives)}}{2 * \textit{(No. of True Positives)} + \textit{(No. of False Positives + No. of False Negatives)}}. \quad (12.4)$$

Matthew correlation coefficient (MCC) is another measure that provides an excellent tradeoff between optimal positive and negative accuracy and is given as

$$\textit{MCC} = \frac{A}{\sqrt{B*C*D*E}} \quad (12.5)$$

$$A = \textit{(No. of True Positives * No. of True Negatives)} - \textit{(No. of False Positives * No. of False Negatives)}$$

$$B = \textit{(No. of True Positives + No. of False Positives)}$$

$$C = \textit{(No. of True Positives + No. of False Negatives)}$$

$$D = \textit{(No. of True Negatives + No. of False Positives)}$$

$$E = \textit{(No. of True Negatives + No. of False Negatives)}$$

MCC score ranges from –1 to 1, where –1 indicates the worst classification performance and 1 is the most accurate classification performance. MCC is the most suitable measure of performance for the imbalance dataset cases as it provides the best tradeoff between positive and negative accuracies.

12.4 Classification of Feature Selection Methods

Several reviews are available for feature selection in Chemo & Bioinformatics [3–9]. The three major attribute selection methods are 1) filter ranking, 2) wrapper methods, and 3) embedded algorithms. Filter ranking is facilitated by different heuristic techniques. The ranking is provided by evaluating the intrinsic properties of the attributes and their correlation with the output. Once a ranking is established, the best subset of the top-ranked attributes can be quickly found out. These methods can provide a ranking without repeated use of classifiers, and, hence, they are fast and have very little computational complexity. The filter ranking, however, may not be very accurate. Filter ranking hybridized with wrapper methods can provide superior performance. Filter methods are fast and need not repeatedly employ a classifier for ranking features. They are, however, not very accurate. Wrapper methods, on the other hand, are quite accurate but rather slow and require large computational efforts. They need to repeatedly use a classifier for the evaluation and ranking of features. Wrapper methods can be classified as deterministic methods or stochastic methods. In embedded techniques, the classification model itself facilitates attribute selection. RF algorithm has two different attribute ranking methods embedded in it. These are known as mean decrease in Gini importance measure and mean decrease in accuracy measure. SVMs also have an embedded attribute ranking technique which is known as SVM recursive feature elimination (SVM-RFE) technique. SVM-RFE recursively eliminates the feature having the least absolute weight value. With this technique, the number of runs for rank estimation is equivalent to the total number of attributes in the data set.

12.4.1 Filter Ranking Methods

Several filter ranking methods are available in the literature. Each of these methods depends on a specific heuristic. With these heuristics, the attributes can be ranked. Once the ranking is established, algorithm performance can be measured, leaving one attribute at a time. Once this is done, the top-ranking attribute subset with maximal algorithm performance can be identified. A final model can be built with this subset, which can be used for testing new examples (Figure 12.1). We discuss a few of the more important filter ranking methods.

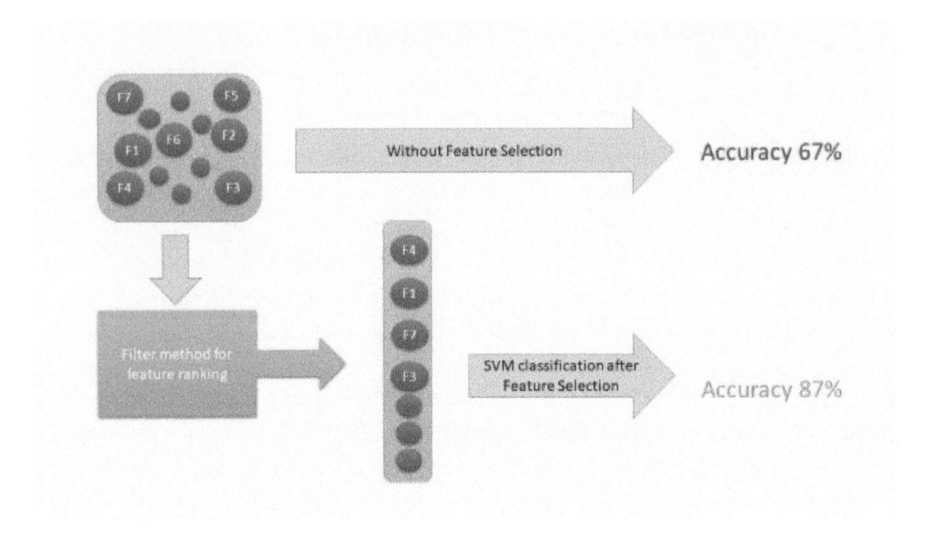

Figure 12.1 Illustration of the filter method.

12.4.2 Chi-Square Test

This filter is used to compute for each input attribute a χ^2 test of independence between the response variable and that attribute. The resulting value of the χ^2 statistics is employed as a quantitative score for ranking.

$$\chi^2\left(X_v\right) = \sum_{a=1}^{m}\sum_{b=1}^{n}\frac{(o_{ab}^{(v)} - e_{ab}^{v})^2}{e_{ab}^{(v)}} \tag{12.6}$$

where $o_{ab}^{(v)}$ represents the observed number of examples with class a, $a \in \{1,\ldots,m\}$, and a value of X_v of the $b$thbin, $b \in \{1,\ldots,n\}$. $e_{ab}^{(v)} = \frac{1}{n}\sum_{a=1}^{m} o_{ab}^{(v)} \cdot \sum_{b=1}^{n} o_{ab}^{(v)}$ is the expected number of examples with class a and a value of X_v of the $b$thbin. The higher the value of this score, the higher is the correlation between that output label and the attribute. Appropriate discretization methods can be used to discretize continuous features.

12.4.3 Fisher's Score

Fisher score is a very popular filter ranking method and one of the most widely used supervised feature selection methods. It is defined as

$$X_v = \frac{\sum n_j(\mu_{ij}-\mu_i)^2}{\sum n_j * \sigma_{ij}^2} \tag{12.7}$$

where i and j denote feature and class, respectively, μ and σ denote mean and the variance, respectively, and n is the number of instances. Attributes with higher value of Fisher scores are ranked higher. With the ranking, it would be computationally easy to select the subset of features that maximize the performance of a given classifier.

12.4.4 Correlation Coefficient

Univariate linear correlation of a feature with the output variable provides a measure of importance and information content of that feature for the classification problem. While a higher correlation between a feature and output is highly desirable, the input features must be uncorrelated. The Pearson correlation of a feature with the output can be computed as

$$\textit{Correlation Coefficient} = \frac{\sum (x_i - \bar{x})(y_i - \bar{x})}{\sqrt{\sum (x_i - \bar{x})^2 \sum (y_i - \bar{y})^2}} \tag{12.8}$$

12.4.5 Mean Absolute Difference

Mean absolute difference is a simple but elegant way of quantifying the importance of a given attribute. It computes the absolute deviation from the mean. The predictive power of an attribute increases with an increase in the mean absolute difference.

12.4.6 Dispersion Ratio

Dispersion ratio filter computes geometric mean and arithmetic mean. For a dataset with n class labels, these values can be found as

$$\textit{Arithmetic Mean} = A_i = \bar{X}_i = \frac{1}{n}\sum_{j=1}^{n} X_{ij}, \tag{12.9}$$

$$\textit{Geometric Mean} = G_i = \left(\prod_{j=1}^{n} X_{ij}\right)^{\frac{1}{n}} \tag{12.10}$$

respectively; since $A_i \geq G_i$, with equality holding if and only if $X_{i1} = X_{i2} = \ldots = X_{in}$, then the ratio

$$R_i = \frac{A_i}{G_i} \tag{12.11}$$

can be used as a dispersion measure. Higher dispersion corresponds to higher discriminative power. The constraints in the above method are that the feature values have to be positive. This can be achieved by normalization.

12.4.7 Variance

Variance is another simple but effective measure to gauge the discriminative power of an attribute:

$$\textit{Variance of } (X) = \frac{\sum (x_j - \bar{x})^2}{n-1.} \tag{12.12}$$

The score helps in removing features with small values of variance which can be due to noise and have very little discriminative power. To use this filter, it is required to scale the data to the same range and without unit variance normalization.

12.4.8 Mutual Information

The entropy of the output variable Y can be defined as

$$H(Y) = -\sum_{y} p(y)\log_2(p(y)) \tag{12.13}$$

And the conditional entropy of Y given X is given by

$$H(Y|X) = \sum_{y} p(y)H(Y|X = x) = \sum_{x} p(x)\left(-\sum_{y} p\,(y|x)\log_2(p(y|x))\right) \tag{12.14}$$

The mutual information of two variables can now be defined as

$$I(Y;X) = H(Y) - H(Y|X). \tag{12.15}$$

This quantity represents the mutual information shared by the two variables. To use this filter ranking measure, the continuous features must be appropriately discretized.

12.4.9 Wrapper Methods

Filter methods are computationally faster and need not repeatedly employ a classifier for ranking features. They are, however, not very accurate. Wrapper methods, on the other hand, are quite accurate but rather slow and require

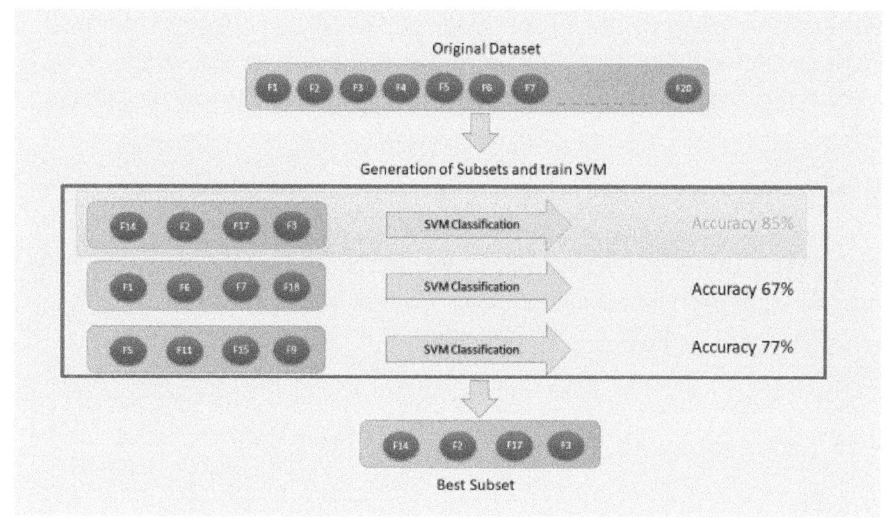

Figure 12.2 Illustration of the wrapper method.

large computational efforts. They need to repeatedly use a classifier for evaluation and ranking of features (Figure 12.2). Wrapper methods can be classified as deterministic or stochastic methods. We describe different deterministic and stochastic wrapper feature selection methods in detail in the following subsections.

12.4.10 Deterministic Wrappers

12.4.10.1 Forward selection

In this method, we first evaluate the performance measure of a classifier with every single feature as input and determine the feature providing the highest performance. This feature is the ranked highest, and in the next step, along with this feature, one more feature is added in each classifier, and every "two feature" combination is evaluated. The best two-feature combination is found in this process and simultaneously two top-ranking features are obtained. The process is repeated, and, each time, an additional feature is ranked. This procedure is completed when all the features are ranked. Once the entire dataset is added to the list, we can determine the best subset of features.

12.4.10.2 Backward selection

In this method, we initiate the algorithm with the entire set of features and evaluate the performance measure. We remove one feature at a time and evaluate the performance of the reduced datasets. The feature causing the least reduction in performance or highest increase in performance is removed at this stage. This process is repeated until we end up with the empty dataset.

12.4.10.3 Bi-directional elimination (stepwise selection)

It is similar to forward selection methodology, but the difference is that, during the addition of a new feature, it also verifies the presence of insignificant features added at every stage and removes them by backward selection. This selection method is a judicious combination of both forward and backward selections.

12.4.11 Stochastic and Heuristic Wrapper Methods

Many wrapper algorithms employ heuristic and stochastic optimization techniques for feature selection. These include genetic algorithms (GA), ant colony optimization (ACO) and other warm-based methods, biogeography based algorithms, group search optimizers, black hole optimization, equilibrium optimizer, etc. We provide a detailed discussion of three of the most popular algorithms in this chapter.

12.4.11.1 Genetic algorithm

GA [10] is a very popular stochastic optimization method for solving feature selection problems. This method mimics the natural selection mechanism in genetics and biological evolution. In nature, genes evolve over successive generations with a view to adapt themselves to the surroundings. This natural phenomenon is successfully used by GA for solving several optimization problems. In Chemo & Bioinformatics domain, GA has been successfully employed for selecting the most informative features. GA feature selection algorithm (Figure 12.3) starts with a randomly initiated population of trial solutions in the coded representation known as chromosomes. The trial solutions comprise randomly generated subsets of features. For the purpose of attribute selection, the commonly employed binary representation (Figure 12.4) consists of zeros and ones for the chromosomes. Every chromosome in the population contains the number of bits equivalent to the number of attributes selected in that chromosome. For every population member, for each bit, zero or one is randomly filled. A bit or gene filled with one represents a feature is selected and a bit filled with zero represents a feature is not selected. With this representation, every population

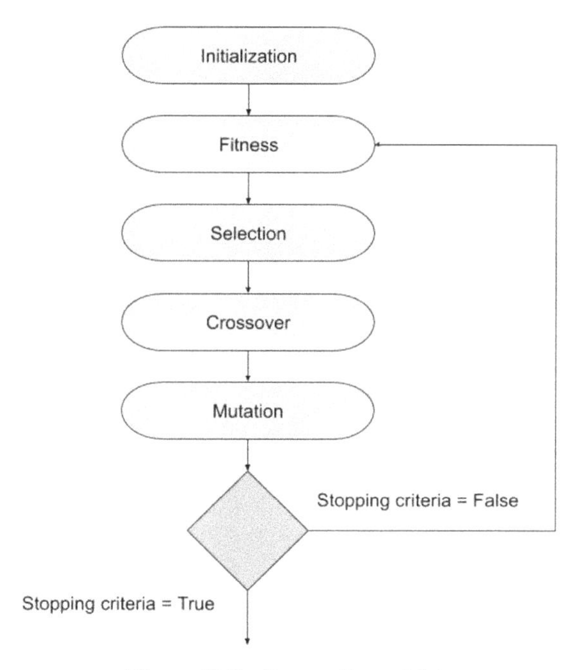

Figure 12.3 Process flow of GA.

in the coded form contains a specific subset of features. After generating the population member, the fitness of each member is evaluated. For this, the dataset with the subset of features represented by every individual is classified with a robust classification method like SVM and an appropriate measure like accuracy is evaluated. After evaluation, the fitness of the solution is evaluated. Subsequent to fitness evaluation, in each generation, three main operations known as selection, crossover, and mutation are conducted to improve the quality of solutions, and, after a few generations subject to a stopping criterion, the algorithm attains near optimal solutions. The procedure is repeated, and the algorithm parameters are tuned to get the best possible solution subject to a previously determined performance measure. The performance measure can be total accuracy, positive accuracy (sensitivity), negative accuracy (specificity), F-measure, MCC, or receiver operating characteristics (ROC).

After fitness is assigned, there is a need to select individuals. Two different selection operators are normally used. The first scheme is known as tournament selection as shown in Figure 12.5. In this method, two individuals are randomly selected, and the one with better fitness is selected.

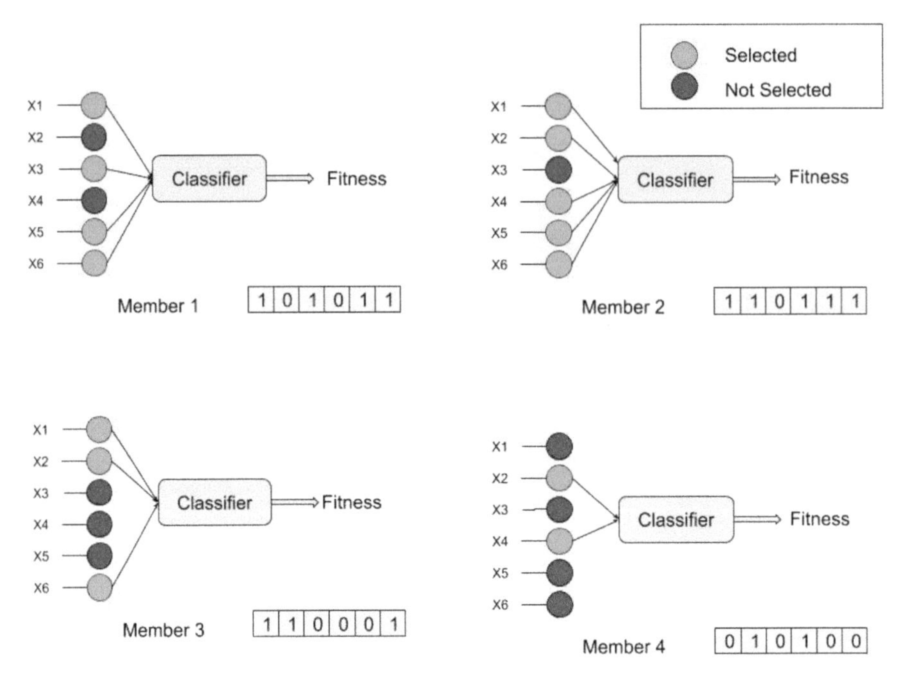

Figure 12.4 Representation of features in GA.

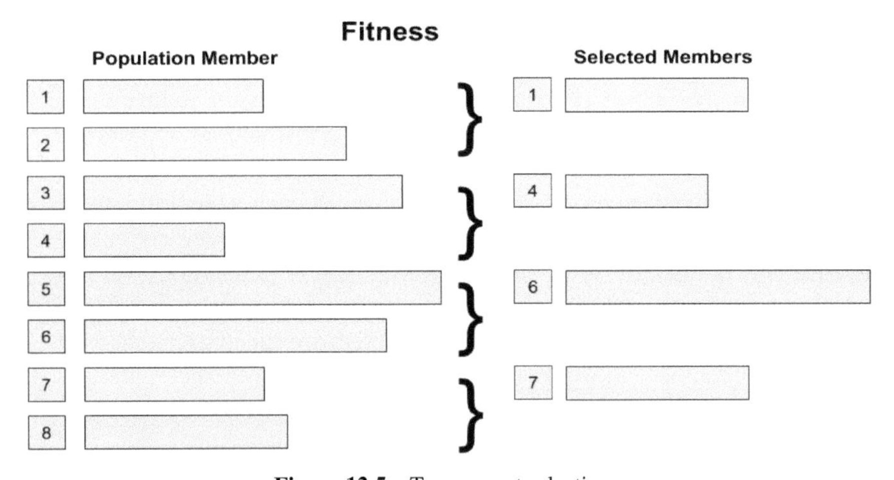

Figure 12.5 Tournament selection.

This process is repeated till all individuals undergo a selection procedure. The process is repeated for one more cycle. In the selection procedure, the number of individuals does not increase or decrease, but copies of fitter individuals may be more than one and some individuals with lesser fitness may get filtered out. In another popular selection method, viz., Roulette wheel selection (Figure 12.6), the population members occupy a space in the wheel proportional to their fitness and the wheel is rotated "N" (where N is the population size) times, and, each time, the individual on which the pointer settles is selected. After selection, the crossover operation is conducted. In this operation, two individuals are randomly selected, and the crossover mechanism is conducted with a probability proportional to a predetermined crossover probability. As shown in Figure 12.7, a common partition in the parent chromosome is randomly selected and the first child is formed by fusing the first segment (section till random index) of the first parent and the second segment of the second parent. The second child is formed by the combination of the first segment (remaining section) of the second parent and the second segment of the first parent. This crossover operation is carried out in a similar fashion with other members until the population size after crossover is the same as the population size before crossover. Subsequent to the crossover operation, the mutation operation is carried out. The mutation step helps in avoiding rapid convergence to poor local optima as it can take the population to any part of the landscape. In the bit flip mutation operation, a bit with a value of one is flipped to zero and vice versa. The iterations are repeated for several generations until a previously defined stopping criterion

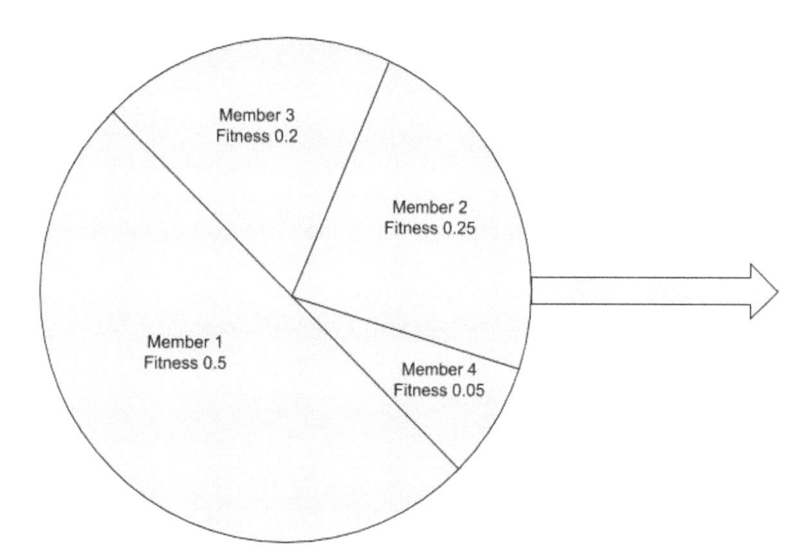

Figure 12.6 Roulette wheel for selection.

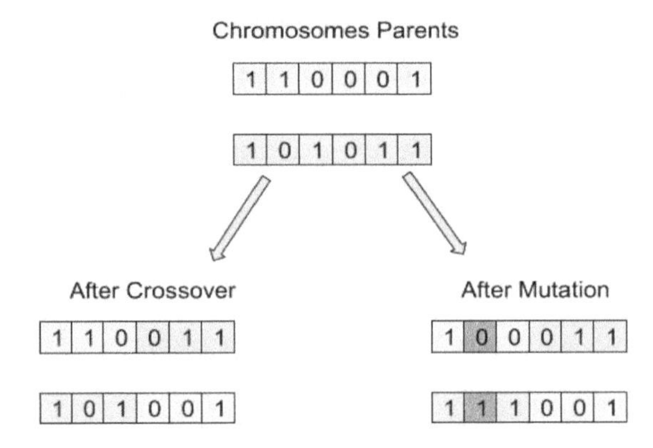

Figure 12.7 Crossover and mutation illustration.

is satisfied. Usually, a certain number of generations are carried out and the fitness is checked. Again, the GA operations are carried out on 5% of the total number of generations and if the improvement is less than a certain predefined value, the operations are stopped.

12.4.11.2 Ant colony optimization

ACO [11] is inspired by the real-life behavior of ants. When ants leave their nest in search of food and reach the food source, they deposit a chemical known as pheromone on the route. Ants not only deposit pheromones but also get attracted to them. They use this as an indirect means of communication, and it can be proven that this autocatalytic feedback mechanism of sensing and getting attracted to pheromone trails can result in the establishment of the shortest route. In the ACO technique for feature selection, features are considered as nodes of a graph with links connecting each node to every other node. Initially, random amounts of pheromone are deposited on all links. Algorithms mainly employ two processes, viz., exploration and exploitation for the selection of the nodes from the current position. At every node, one of these processes is selected by a probability. This probability q_0 is fixed *a priori*. Ants start from a random node and move to the next node by exploitation with a probability (q_0) and by exploration with a probability ($1 - q_0$). If exploitation is selected, an ant moves to the feature whose link has maximum transition probability. If exploration is selected, it moves to the legally permitted connected node with a probability proportional to the transition probability of the links connected to the current node. The transition probability is defined as a product of two terms composed of pheromone concentration and heuristics. This can be defined as follows:

$$f = \begin{cases} \max\left[\tau(f_{ij})\,\eta\,(f_i)^{\beta}\right] & \textbf{\textit{If}}\ (\textbf{\textit{rnd}} \leq q_0)\ \textbf{\textit{Exploration}} \\ \dfrac{\tau\,(f_{ij})\,\eta\,(f_i)^{\beta}}{\sum \tau(f_{ij})\,\eta\,(f_i)^{\beta}} & \textbf{\textit{Otherwise Exploitation}} \end{cases} \tag{12.16}$$

In this way, an ant travels from one node to another until the completion of the tour. Similarly, different ants conduct their tours. The selected subset of features is evaluated by a robust classifier and the pheromones of links of the ant with maximum performance are enhanced with a quantity proportional to the tour quality. Pheromones of all links are reduced by a small percentage. This pheromone evaporation helps to avoid convergence to a suboptimal solution. The iterations are continued until convergence and the best subset is found. In binary ACO, each attribute is assumed to have one of two states zero and one, hence the name Binary ACO. State one indicates that the feature corresponding to that state is selected, while zero indicates the feature is not selected. If the features represent the nodes of a graph, between any two nodes, four different types of links are possible. Unlike in classic ACO, in this algorithm, ants visit all the attributes. Finally, a given ant after visiting all features outputs a vector of 1s and 0s of length equal to the number of attributes. Recently, Kashef and Nezambadipour [12] employed an advanced version of binary ACO incorporating correlation between features. Their binary algorithm employs four different heuristic methods as filters and hybridizes with ACO wrapper providing excellent performance.

A filter-wrapper illustration is shown below in Figure 12.8.

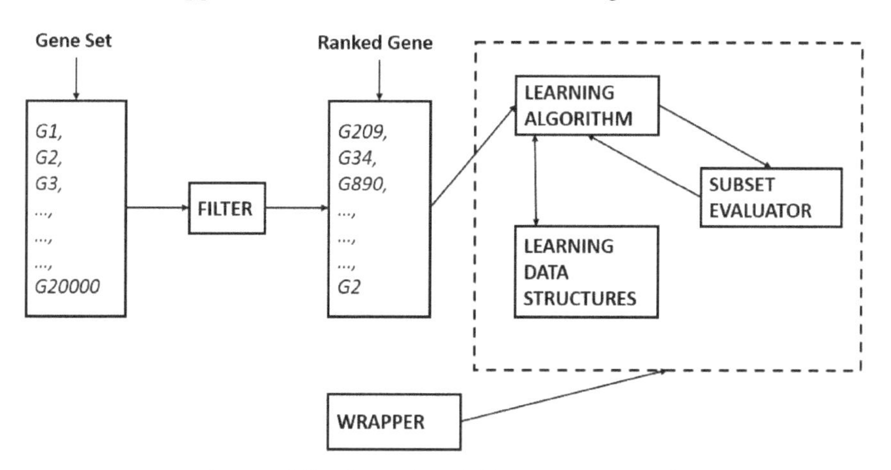

Figure 12.8 Illustration of filter-wrapper method.

12.4.11.3 Biogeography-Based optimization

The biogeography-based optimization (BBO) [13] is inspired by areal-life phenomenon involving the immigration and emigration of the populations in different parts of the planet. Further, the BBO algorithm mimics the geographical distribution, redistribution, and dynamics of many species over a period. The evolution of BBO candidate solutions for feature selection emulates the natural processes of population migration. Such an iterative process results in a small subset of attributes when given high dimensional data with many attributes. BBO has been applied to several domains including drug discovery, disease classification, and medical diagnosis.

12.4.12 Embedded Methods

In embedded methods, it is possible to rank the features during the classification/regression model development process itself. SVMs for classification and regression, RF for classification and regression, and Lasso penalty-based regression are some of the examples of embedded methods. The RF algorithm provides ranking of the feature based on Gini importance and variable importance. The SVM algorithm can be used for RFE (SVM-RFE). We explain these algorithms in detail below.

12.4.12.1 SVM-RFE algorithm

SVM-RFE algorithm for feature ranking has become immensely popular. The algorithm starts with all features; the hyperplane equation employing linear kernel is found and the weights of all attributes are estimated. The algorithm assumes that the attribute weights signify the importance of the respective attributes. Thus, the attribute with the least absolute value contributes the least to the overall hyperplane equation. This attribute is first removed, and the model is rebuilt with the rest of the attributes and the process is repeated until all attributes are ranked. It is then possible to select the top-ranked feature subset.

12.4.12.2 Random forest mean decrease in accuracy

The RF algorithm steps can be explained as follows. 1) run the algorithm with all features and obtain the Out of Bag (OOB) vote count for the correct label. 2) Permute each attribute, one at a time and estimate the OOB vote counts. 3) Find the difference between the two OOB vote counts for each attribute in each tree of the entire forest and estimate the average of this quantity. The estimated quantity represents the mean decrease in accuracy for each attribute. A significant attribute will have a large value for this quantity and will have a higher rank.

12.4.12.3 Random forest mean decrease in Gini

Every time a split is made in a tree, the Gini score value reduces, indicating the increase in purity of the children nodes. This Gini decrease is added for all splits of a given feature in a tree. Similarly, Gini decreases over all splits in the other trees are added. This grand addition provides an indication of the segregation capability of a given feature and can serve as a fast indicator of feature importance.

12.5 Bioinformatics Applications

Alièkoviæ and Subasi [14] have used two different breast cancer datasets. Their two-stage algorithm uses GA in the first step to eliminate insignificant attributes. In the next stage, multiple algorithms are employed for classifying the datasets. Three different measures are used to evaluate algorithm performance. These include ROC, F-measure, and overall accuracy. The combination of the RF and GA model selected 14 features providing an accuracy greater than 99%.

Zhang *et al.* [15] used an adaptive GA with logistic regression classifier for discovering the top-ranking subset of biomarkers for prediction of the Alzheimer's disease progression. Adaptive GA employs probabilistic crossover and mutation operators. This strategy preserves high fitness solutions and disrupts low fitness solutions effectively. The initial set of attributes include 53 clinical pathological measures, 7 measures of circulating metals, 111 protein features, 7 plasma measures, and 3 demographics features. The top-ranked 20–30 GA features were able to achieve an area under the curve (AUC) between 0.83 and 89.

Soufan *et al.* [16] have developed web based wrapper feature selection tool. This tool is a parallel GA implementation, which is robust and facilitates attribute for many bioinformatics problems. The web based wrapper feature selection tool also hybridizes various filter ranking heuristics to improve algorithm performance. The tool was executed on different biomedical datasets. This tool produces a small subset of high-performance features yielding excellent accuracies.

ACO has been found to be a very efficient and versatile methodology for solving attribute selection problems in a variety of situations in Chemo & Bioinformatics domains. Gene selection is an important application in which ACO attribute and biomarker selection is found to perform extremely well. Microarray datasets in disease prediction have very high dimensionality with large number of gene expression profiles as attributes. Most of these are irrelevant, noisy, and provide less information. The difficult task of obtaining a small set of attributes from this high dimensional data is a very difficult task handled exceptionally well by the ACO algorithm. The reduced biomarker genes have been found to have extremely high predictive power, simultaneously providing valuable domain knowledge. These methods were initially developed and discussed in detail in [17, 18].

In [19], ACO was used with the SVM classifier for gene selection. The experiments were conducted on colon cancer, brain cancer, and leukemia datasets. The authors applied certain preprocessing techniques on these datasets before applying them for gene selection. This involves discretization of the datasets, application of various filters that removed genes having little variation across samples, and removal of genes that showed values less than a threshold for a ratio of maximum to minimum gene expression levels computed for each gene. Parameters used for ACO-SVM also had to be tuned to ensure the highest performance. For SVM, linear kernels were tested against polynomial and Radial Basis Function (RBF) kernels. For this implementation, ACO used 40 ants and 40 generations for the experiments. Extensive simulations were done for subset sizes ranging from 20 to 100. A leave one out cross-validation (LOOCV) approach was used for evaluating the gene subsets selected by this method. They obtained 100% accuracy with their ACO feature selection approach.

Xiong and Wang [20] proposed a hybrid ACO-based technique toward gene selection and cancer classification. They used the *t*-statistic and variable importance measures from RF as a heuristic to pre-select genes using ACO and send the selected gene subset to the RF and retrieve the classification accuracy which ACO used for later iterations. As for post-processing in each generation, the algorithm also used a forward selection method for improved gene subset prediction. For the colon dataset [21] with a reduced subset size of four features, they obtained an accuracy of 96.77%, and for the Leukemia dataset [22] with a reduced subset size of three features, they obtained an accuracy of 100%.

The affinity of binding between peptides and Major Histocompatibility Complex Class I (MHC-I) protein molecules is a vital task in viral biology and vaccine design. Srivatsava *et al.* [23] have developed a BBO-based hybrid filter-wrapper algorithm for these prediction tasks. They employed both RF and SVM as classifiers. They carried out simulations with Comparative Evaluation of Prediction Algorithms (CoEPrA) (available at http://www.coepra.org). They obtained excellent accuracies for these datasets reduced with the BBO wrapper algorithm.

Wang *et al.* [24] developed an Alzheimer disease detection system with better performance than existing methods. Twenty-eight diseases and 98 negative images were shortlisted from the OASIS dataset. They combined BBO, wavelet entropy, and multilayer perceptron and obtained an overall accuracy of over 92%.

Thawkar *et al.* [25] employed BBO for the identification of breast cancer. The dataset consisted of 651 mammograms. From the extracted intensity, texture, and shape-based features, these authors selected the most informative ones with BBO. These selected features were then classified with a combination of neuro-fuzzy inferencing and neural nets; these authors concluded that such a synergistic combination produced excellent results. They further obtained an accuracy greater than 98% and equally good specificity, sensitivity, and ROC measures. Some of these feature selection applications are tabulated in Table 12.1.

12.5.1 Example Involving Quantitative Structure Activity Relationship Application

Rapid evaluation and shortlisting of potentially useful small molecules can be accomplished by quantitative structure activity relationship (QSAR) models. QSAR can rapidly evaluate and shortlist potentially important small molecules; these models can further reduce experimental screening and are very computationally efficient, effective and robust tools. Given a set of compounds exhibiting known activities for a given medical task, various types of descriptors can be extracted. These include electronic, geometrical, quantum chemical, topological, and descriptors of various dimensions. Not all descriptors may contain relevant information, and, hence, obtaining a subset of the most important attributes is again a very key step in QSAR modeling. The following steps need to be carried out in the QSAR modeling procedure. The QSAR modeling steps include 1) getting the descriptors of different types using domain information if available, 2) employing appropriate selection methods to shortlist a small subset of these, 3) building classification/regression models, 4) discriminating among different models and screening and shortlisting a set of activities. Both regression and classification models can be built. Regression models predict actual activities, whereas classification models make binary predictions; output whether the molecule is active or inactive. For building a classification model, we first need a proper threshold based on prior knowledge to divide the training set of compounds as active and inactive. Excellent reviews on descriptors selection for QSAR modeling are presented by [26–29].

Table 12.1 Literature Review of Bioinformatics Application.

References	Feature Selection Method	Dataset and Features Used	Classifier(s)	Performance Measure
Lan and Vucetic, 2011 [30]	Filter ranking employing multi-task method	Multi-tissue using microarray expression profile features	Regression with lasso penalty; Logistic regression including penalization; Support vector machines	Classification accuracy; Kruskal-Wallis test
Xiang *et al.*, 2012 [31]	Discriminative least square regression (DLSR)	Glioma; Lung cancer; Mixed lineage leukemia cancer; SRBCT; CLL-SUB-111; GLA-BRA-180; TOX-171 dataset with expression profile attributes	Nearest neighbor classifier	10-fold cross-validation; Complexity measure; Students' t tests
Ben-Dor *et al.*, 2000 [32]	Clustering-based classification using CAST (cluster affinity search technique) algorithm	Colon; Ovarian; Leukemia using microarray expression profile features	SVM; Nearest neighbor classifier (NNC); Boosting	Leave one out cross-validation (LOOCV); ROC curves
Chuang *et al.*, 2009 [33]	Particle Swarm Optimization (PSO) wrapper with information gain filter ranking	11_Tumors dataset; Brain_Tumor 1 dataset; Leukemia 2 dataset; 9_Tumors dataset; Leukemia1 dataset; 14_Tumors dataset; Prostate_Tumor dataset; SRBCT dataset;	k-nearest neighbors (KNN); SVM	Fitness function; Classification accuracy

Table 12.1 (Continued)

References	Feature Selection Method	Dataset and Features Used	Classifier(s)	Performance Measure
		Diffuse large B-cell lymphomas (DLBCL) dataset; Brain_Tumor 2 dataset; Lung cancer dataset. Average using microarray expression profile features		
Mao and Yang, 2019 [34]	Multilayer wrapper feature selection with randomized search strategy	Breast using 7129; Colon using 2000; CNS using 7129; Hepatoma 7129. Leukemia 7129 gene expression profile features	SVM	Classification accuracy
Dudoit and Fridlyand, 2002 [35]	Cluster classification using Partitioning Around Medoids (PAM) based on Euclidean distance metric	Lymphoma; Leukemia; NCI 60; Melanoma using microarray expression profile features	Diagonal linear discriminant analysis (DLDA)	Prediction accuracy
Yeung and Bumgarner, 2003 [36]	Uncorrelated shrunken centroid with and without error weighting	Multiple tumors; Breast using microarray expression profile features	SVM classifier	Classification accuracy; No. of bio marker genes; Stability measure of features
Inbarani *et al.*, 2014 [37]	Supervised PSO-based relative reduct	Erythemato-squamous diseases using 12 clinical and 22 histopathological features;	Naive Bayes; Bayes net; KStar	Precision; Recall; Overall accuracy

(Continued)

Table 12.1 (Continued)

References	Feature Selection Method	Dataset and Features Used	Classifier(s)	Performance Measure
		Breast (diagnostic and prognostic) using 30 features describing characteristics of cell nuclei and one class attribute; single-photon emission computed tomography features (SPECTF) using single-photon emission computed tomography (SPECT) image features resulting in 44 continuous feature patterns		
Fortino *et al.*, 2014 [38]	Fuzzy pattern	Multiple datasets including: Leukaemia; Lung; Psoriasis; Peripheral Blood Mononuclear Cells (PBMC) multiple feature sets	RF-based classifier	F-score; G-mean metric; Feature stability
Gevaert *et al.*, 2006 [39]	Bayesian network learning	Expression profile attribute, age, tumor diameter, grade details of tumor, estrogen and progesterone receptor status, absence or presence of angioinvasion, and lymphocytic invasion	Least squares SVM	ROC
Bhanot *et al.*, 2006 [40]	Robust feature selection using peak extraction, filtering and support set selection	Prostate using mass spectra features	Support vector machines; Neural networks; Ensemble weighted voting;	Sensitivity analysis; Specificity analysis

Table 12.1 (Continued)

References	Feature Selection Method	Dataset and Features Used	Classifier(s)	Performance Measure
			KNN; Decision trees classifier; Logistic regression classification	
Wang *et al.*, 2007 [41]	Parsimonious threshold-independent attribute selection	Prostate; Liver using mass spectra features	Least angle regression (LARS) method	ROC curve
Kim *et al.*, 2010 [42]	Embedded attribute selection with multi-layer perceptron	HIV-1 protease cleavage dataset 14 features selected from 160 original features	Multilayer Perceptron (MLP); Decomposition approach of an OAS algorithm	F-measure
Kavakiotis *et al.*, 2017 [43]	Frequent item feature selection (FIFS)	Single Nucleiotide Polymorphism (SNP) dataset of various pig breeds (14 subclasses) prevalent in the UK	Bayesian maximum likelihood assignment criteria	Pairwise Wright's fixation index (FST); Informativeness for assignment (Delta value)
Peng *et al.*, 2010 [44]	Hybrid method using pre-selection filter, wrapper of SVM, and sequential forward floating search (SFFS)	Breast (diagnostic and prognostic) with more than 32 and 34 features, respectively SPECTF heart with 44 features; Microcalcification detection with 39 features	SVM;	Classification accuracy; *k*-fold cross-validation and ROC
Saeys *et al.*, 2004 [45]	Estimation of distribution algorithm (EDA)	Splice sites for arabidopsis thaliana with Dataset 1 containing 400	Naive Bayes; SVM	F-measure

(Continued)

Table 12.1 (Continued)

References	Feature Selection Method	Dataset and Features Used	Classifier(s)	Performance Measure
		position-dependent binary features; Dataset 2 containing 400 position-dependent and 128 position-independent binary features; Dataset 3 containing 528 binary features from 1 and 2 and 1568 position-dependent dimer binary features		
Soto *et al.*, 2008 [46]	Genetic algorithm based method	Molecular weight, count of hydrogen atoms, count of alcohols, sum of atomic electronegativities, etc.	Linear regression; Non-linear regression	Decision trees; mean absolute error (MAE); Analysis of variance (ANOVA); Nested ANOVA; Neural Network Ensemble (NNE)
Wang *et al.*, 2006 [47]	Relief-F feature selection algorithm	DNA copy number in SNP array datasets	KNN; SVM; Naive Bayes	LOOCV
Varshavsky *et al.*, 2006 [48]	Singular Value Decomposition (SVD) entropy with forward and backward selection	Leukaemia using microarray expression profile features. Mixed Lineage Leukemia (MLL) using microarray expression profile features;	QC clustering	Jaccard score

Table 12.1 (Continued)

References	Feature Selection Method	Dataset and Features Used	Classifier(s)	Performance Measure
		Viruses dataset containing 18 amino acid composition measurements as features		
Luss and Aspremont, 2010 [49]	Sparse principal component analysis (SPCA)	Colon cancer data; gene expression	Sparse principal component and clustering	Rand index
Ghorai *et al.*, 2011[50]	Minimum redundancy, maximum-relevancy (mRMR) filter	Leukaemia; Colon; Lung; Breast; DLBCL; Prostate; lymphoma, liver	Non-parallel proximal classifier ensemble	k-fold CV (k=10); Majority voting of ensembles
Sahlol *et al.*, 2020 [51]	Hybrid fractional order marine predator's algorithm	Twochest X-ray datasets for COVID-19	Convolutional neural networks	F-score; Fitness values
Zainuddin *et al.*, 2019 [52]	ACO	Colon cancer and lung cancer dataset	Fuzzy inference based on adaptive network	Accuracy
Li *et al.*, 2018 [53]	SVM-RFE (support vector machine based on recursive feature elimination)	Sixgene expression microarray datasets using microarray expression features; Leukaemia (2000 features). Colon (7129 features); Prostate (7129 features); Ovarian (15,154 features); Breast (12,600 features); CNS (24,481 features)	SVM; Naïve Bayes; KNN; Logistic regression (LR);	Accuracy (ACC); Area under ROC curve (AUC); Matthew's correlation coefficient (MCC)

(Continued)

Table 12.1 (Continued)

References	Feature Selection Method	Dataset and Features Used	Classifier(s)	Performance Measure
Shi *et al.*, 2011 [54]	*k*-Top scoring pair (*k*-TSP) algorithm	Cancer prognostic datasets using gene expression features	kTSP+ SVM	LOOCV
Malan & Sharma, 2019 [55]	Neighborhood component analysis (NCA)	Brain computer interface (BCI) Competition II, Dataset III, and BCI Competition IV Dataset. Using features of EEG (statistical, frequency, and phase)	SVM	Average accuracy; Kappa coefficient
Parvandeh *et al.*, 2020 [56]	Relief-based feature selection	RNA seq data for Major depressive disorder (MDD) with 15231 genes using expression and quality features	Random forest	Consensus nested cross validation
Deraeve & Alexander, 2018 [57]	Iterative neural network with cross-validation	Simulated dataset; Zeithamova dataset; Haxby dataset	SVM	CV; Jaccard index
Deviaene *et al.*, 2019 [58]	RF feature selection method	Detection of sleep apnea using feature of SpO2 signals in; Sleep heart health study (SHHS) which contains polysomnographs (PSGs) of 5793 general population subjects at baseline (SHHS1) and a follow-up PSG on average 5 years later for 2651 of these subjects (SHHS2)	Random forest	Cohen kappa value
Han, 2017 [59]	Non-negative singular value approximation (NSVA)	RNA seq datasets; Marioni data; Prostate data; Fly embryos data with gene features	Gene contribution scores	Modified Fischer exact test

Table 12.1 (Continued)

References	Feature Selection Method	Dataset and Features Used	Classifier(s)	Performance Measure
Alièkoviæ and Subasi, 2019 [14]	Genetic algorithms	Breast cancer datasets,: Expression profiles	Multiple classifiers	Accuracy, receiver operating characteristics (ROC), F-measure
Zhang *et al.*, 2019 [15]	Adaptive genetic algorithms	Dataset including multiple features 53 clinical pathological measures, 7 measures of circulating metals, 111 protein features, 7 plasma measures, and 3 demographics features.	Logistic regression	Area under the curve (AUC)
Soufan *et al.*, 2015 [16]	Parallel genetic algorithms	Web-based tool tested on biomedical datasets	Multiple classifiers	Multiple performance
Gupta *et al.*, 2007 [19]	ACO	Colon, cancer, and breast cancer, gene expression profiles	SVM	Accuracy
Xiong and Wang, 2009 [20]	ACO	Colon, cancer, and leukemia datasets, gene expression profiles	Random forest	Accuracy
Srivastava *et al.*, 2013 [23]	Biogeography-based optimization	Four different CoEPrA datasets from www.coepra.org, MHC-I binding/non-binding datasets peptides	SVM; Random forests	Classification Accuracy
Wang *et al.*, 2018 [24]	Biogeography-based optimization	Selected examples from OASIS dataset mammogram features	Multilayer perceptron	Classification accuracy
Thawkar *et al.*, 2018 [25]	Biogeography based optimization	651 mammograms; intensity, texture and shape based features	Adaptive fuzzy neuro inference: Artificial neural networks	Accuracy, sensitivity and selectivity

12.6 Final Remarks

Attribute selection is a very essential and useful preprocessing step in diverse applications in the broad areas of Chemo & Bioinformatics domains. In this chapter, we have described different feature selection methodologies in detail. Several illustrative examples have also been provided. This review will be very useful for researchers, academicians, and practicing professionals.

References

[1] Vapnik VN. An overview of statistical learning theory. IEEE transactions on neural networks. 1999 Sep; 10(5):988–99.

[2] Breiman L. Random forests. Machine learning. 2001 Oct 1; 45(1):5–32.

[3] Wang L, Wang Y, Chang Q. Feature selection methods for big data bioinformatics: A survey from the search perspective. Methods. 2016 Dec 1; 111:21–31.

[4] Remeseiro López B, Bolon Canedo V. A review of feature selection methods in medical applications. Computers in Biology and Medicine, 112. 2019.

[5] Belgiu M, Drãguþ L. Random forest in remote sensing: A review of applications and future directions. ISPRS Journal of Photogrammetry and Remote Sensing. 2016 Apr 1; 114:24–31.

[6] Liang S, Ma A, Yang S, Wang Y, Ma Q. A review of matched-pairs feature selection methods for gene expression data analysis. Computational and structural biotechnology journal. 2018 Jan 1; 16:88–97.

[7] Khan PM, Roy K. Current approaches for choosing feature selection and learning algorithms in quantitative structure–activity relationships (QSAR). Expert opinion on drug discovery. 2018 Dec 2; 13(12): 1075–89.

[8] Saeys Y, Inza I, Larrañaga P. A review of feature selection techniques in bioinformatics. Bioinformatics. 2007 Oct 1; 23(19):2507–17.

[9] Tian S, Wang C, Wang B. Incorporating pathway information into feature selection towards better performed gene signatures. BioMed research international. 2019 Apr 3; 2019.

[10] Goldberg DE, Holland JH. Genetic algorithms and machine learning.

[11] Dorigo M, Birattari M, Stutzle T. Ant colony optimization. IEEE computational intelligence magazine. 2006 Nov; 1(4):28–39.

[12] Kashef S, Nezamabadi-pour H. An advanced ACO algorithm for feature subset selection. Neurocomputing. 2015 Jan 5; 147:271–9.

[13] Simon D. Biogeography-based optimization. IEEE transactions on evolutionary computation. 2008 Mar 21; 12(6):702–13.
[14] Alièkoviæ E, Subasi A. Breast cancer diagnosis using GA feature selection and Rotation Forest. Neural Computing and Applications. 2017 Apr 1; 28(4):753–63.
[15] Zhang P, West NP, Chen PY, Thang MW, Price G, Cripps AW, Cox AJ. Selection of microbial biomarkers with genetic algorithm and principal component analysis. BMC bioinformatics. 2019 Dec 1; 20(6):413
[16] Soufan O, Kleftogiannis D, Kalnis P, Bajic VB. DWFS: a wrapper feature selection tool based on a parallel genetic algorithm. PloS one. 2015 Feb 26; 10(2):e0117988.
[17] Sivagaminathan RK, Ramakrishnan S. A hybrid approach for feature subset selection using neural networks and ant colony optimization. Expert systems with applications. 2007 Jul 1; 33(1):49–60.
[18] Robbins KR, Zhang W, Bertrand JK, Rekaya R. The ant colony algorithm for feature selection in high-dimension gene expression data for disease classification. Mathematical Medicine and Biology: a Journal of the IMA. 2007 Dec 1; 24(4):413–26.
[19] Gupta A, Jayaraman VK, Kulkarni BD. Feature selection for cancer classification using ant colony optimization and support vector machines. In Analysis of biological data: a soft computing approach 2007 (pp. 259–280).
[20] Xiong W, Wang C. A hybrid improved ant colony optimization and random forests feature selection method for microarray data. In 2009 Fifth International Joint Conference on INC, IMS and IDC 2009 Aug 25 (pp. 559–563). IEEE.
[21] Alon U, Barkai N, Notterman DA, Gish K, Ybarra S, Mack D, Levine AJ. Broad patterns of gene expression revealed by clustering analysis of tumor and normal colon tissues probed by oligonucleotide arrays. Proceedings of the National Academy of Sciences. 1999 Jun 8; 96(12):6745–50.
[22] Golub TR, Slonim DK, Tamayo P, Huard C, Gaasenbeek M, Mesirov JP, Coller H, Loh ML, Downing JR, Caligiuri MA, Bloomfield CD. Molecular classification of cancer: class discovery and class prediction by gene expression monitoring. science. 1999 Oct 15; 286(5439):531–7.
[23] Srivastava A, Ghosh S, Anantharaman N, Jayaraman VK. Hybrid biogeography based simultaneous feature selection and MHC class I peptide binding prediction using support vector machines and random

forests. Journal of Immunological Methods. 2013 Jan 31; 387(1-2): 284–92

[24] Wang SH, Zhang Y, Li YJ, Jia WJ, Liu FY, Yang MM, Zhang YD. Single slice based detection for Alzheimer's disease via wavelet entropy and multilayer perceptron trained by biogeography-based optimization. Multimedia Tools and Applications. 2018 May 1; 77(9):10393–417.

[25] Thawkar S, Ingolikar R. Classification of masses in digital mammograms using biogeography-based optimization technique. Journal of King Saud University-Computer and Information Sciences. 2018 Feb 1.

[26] Shahlaei M. Descriptor selection methods in quantitative structure–activity relationship studies: a review study. Chemical reviews. 2013 Oct 9; 113(10):8093–103.

[27] Verma J, Khedkar VM, Coutinho EC. 3D-QSAR in drug design-a review. Current topics in medicinal chemistry. 2010 Jan 1; 10(1):95–115.

[28] Khan AU. Descriptors and their selection methods in QSAR analysis: paradigm for drug design. Drug discovery today. 2016 Aug 1; 21(8):1291–302.

[29] Concu R, Cordeiro MN. On the relevance of feature selection algorithms while developing non-linear QSARs. InEcotoxicological QSARs 2020 (pp. 177–194). Humana, New York, NY.

[30] Lan L, Vucetic S. Improving accuracy of microarray classification by a simple multi-task feature selection filter. International journal of data mining and bioinformatics. 2011 Jan 1; 5(2):189–208.

[31] Xiang S, Nie F, Meng G, Pan C, Zhang C. Discriminative least squares regression for multiclass classification and feature selection. IEEE transactions on neural networks and learning systems. 2012 Sep 11; 23(11):1738–54.

[32] Ben-Dor A, Bruhn L, Friedman N, Nachman I, Schummer M, Yakhini Z. Tissue classification with gene expression profiles. In Proceedings of the fourth annual international conference on Computational molecular biology 2000 Apr 8 (pp. 54–64).

[33] Chuang LY, Ke CH, Chang HW, Yang CH. A two-stage feature selection method for gene expression data. OMICS A journal of Integrative Biology. 2009 Apr 1; 13(2):127–37.

[34] Mao Y, Yang Y. A Wrapper Feature Subset Selection Method Based on Randomized Search and Multilayer Structure. BioMed research international. 2019 Nov 4; 2019.

[35] Dudoit S, Fridlyand J. A prediction-based resampling method for estimating the number of clusters in a dataset. Genome biology. 2002 Jun 1; 3(7):research0036–1.

[36] Yeung KY, Bumgarner RE. Multiclass classification of microarray data with repeated measurements: application to cancer. Genome biology. 2003 Dec 1; 4(12):R83.

[37] Inbarani HH, Azar AT, Jothi G. Supervised hybrid feature selection based on PSO and rough sets for medical diagnosis. Computer methods and programs in biomedicine. 2014 Jan 1; 113(1):175–85.

[38] Fortino V, Kinaret P, Fyhrquist N, Alenius H, Greco D. A robust and accurate method for feature selection and prioritization from multi-class OMICs data. PloS one. 2014 Sep 23; 9(9):e107801.

[39] Gevaert O, Smet FD, Timmerman D, Moreau Y, Moor BD. Predicting the prognosis of breast cancer by integrating clinical and microarray data with Bayesian networks. Bioinformatics. 2006 Jul 15; 22(14):e184–90.

[40] Bhanot G, Alexe G, Venkataraghavan B, Levine AJ. A robust meta-classification strategy for cancer detection from MS data. Proteomics. 2006 Jan; 6(2):592–604.

[41] Wang Z, Chang YC, Ying Z, Zhu L, Yang Y. A parsimonious threshold-independent protein feature selection method through the area under receiver operating characteristic curve. Bioinformatics. 2007 Oct 15; 23(20):2788–94.

[42] Kim G, Kim Y, Lim H, Kim H. An MLP-based feature subset selection for HIV-1 protease cleavage site analysis. Artificial intelligence in medicine. 2010 Feb 1; 48(2-3):83–9.

[43] Kavakiotis I, Samaras P, Triantafyllidis A, Vlahavas I. FIFS: A data mining method for informative marker selection in high dimensional population genomic data. Computers in biology and medicine. 2017 Nov 1; 90:146–54.

[44] Peng Y, Wu Z, Jiang J. A novel feature selection approach for biomedical data classification. Journal of Biomedical Informatics. 2010 Feb 1; 43(1):15–23.

[45] Saeys Y, Degroeve S, Aeyels D, Rouzé P, Van de Peer Y. Feature selection for splice site prediction: a new method using EDA-based feature ranking. BMC bioinformatics. 2004 Dec 1; 5(1):64.

[46] Soto AJ, Cecchini RL, Vazquez GE, Ponzoni I. An evolutionary approach for feature selection applied to ADMET prediction. Inteligencia Artificial. Revista Iberoamericana de Inteligencia Artificial. 2008; 12(37):55–63.

[47] Wang Y, Makedon F, Pearlman J. Tumor classification based on DNA copy number aberrations determined using SNP arrays. Oncology reports. 2006 Apr 1; 15(4):1057–9.

[48] Varshavsky R, Gottlieb A, Linial M, Horn D. Novel unsupervised feature filtering of biological data. Bioinformatics. 2006 Jul 15; 22(14):a e507–13.
[49] Luss R, d'Aspremont A. Clustering and feature selection using sparse principal component analysis. Optimization and Engineering. 2010 Feb 1; 11(1):145–57.
[50] Ghorai S, Mukherjee A, Sengupta S, Dutta PK. Cancer classification from gene expression data by NPPC ensemble. IEEE/ACM Transactions on Computational Biology and Bioinformatics. 2010 May 20; 8(3): 659–71.
[51] Sahlol AT, Yousri D, Ewees AA, Al-Qaness MA, Damasevicius R, Abd Elaziz M. COVID-19 image classification using deep features and fractional-order marine predators algorithm. Scientific Reports. 2020 Sep 21; 10(1):1–5.
[52] Zainuddin S, Nhita F, Wisesty UN. Classification of gene expressions of lung cancer and colon tumor using Adaptive-Network-Based Fuzzy Inference System (ANFIS) with Ant Colony Optimization (ACO) as the feature selection. InJournal of Physics: Conference Series 2019 Mar (Vol. 1192, No. 1, p. 012019). IOP Publishing.
[53] Li Z, Xie W, Liu T. Efficient feature selection and classification for microarray data. PloS one. 2018 Aug 20; 13(8):e0202167.
[54] Shi P, Ray S, Zhu Q, Kon MA. Top scoring pairs for feature selection in machine learning and applications to cancer outcome prediction. BMC bioinformatics. 2011 Dec 1; 12(1):375.
[55] Malan NS, Sharma S. Feature selection using regularized neighborhood component analysis to enhance the classification performance of motor imagery signals. Computers in biology and medicine. 2019 Apr 1; 107:118–26.
[56] Parvandeh S, Yeh HW, Paulus MP, McKinney BA. Consensus features nested cross-validation. Bioinformatics. 2020 May 1; 36(10):3093–8.
[57] Deraeve J, Alexander WH. Fast, accurate, and stable feature selection using neural networks. Neuroinformatics. 2018 Apr 1; 16(2):253–68.
[58] Deviaene M, Testelmans D, Borzée P, Buyse B, Van Huffel S, Varon C. Feature selection algorithm based on random forest applied to sleep apnea detection. In2019 41st Annual International Conference of the IEEE Engineering in Medicine and Biology Society (EMBC) 2019 Jul 23 (pp. 2580–2583). IEEE.
[59] Han H. A novel feature selection for RNA-seq analysis. Computational biology and chemistry. 2017 Dec 1; 71:245–57.

13

Bifurcation Analysis for COVID-19 Model With Inhibitory Effect

Nita H. Shah, Nisha Sheoran*, **and Ekta Jayswal**

Department of Mathematics, Gujarat University, Ahmedabad, Gujarat, India
*Corresponding Author
E-mail: nitahshah@gmail.com; sheorannisha@gmail.com;
jayswal.ekta1993@gmail.com

Abstract

Different stages of unlocking have begun for COVID-19 pandemic in some parts of the world. Therefore, it becomes important to focus on inhibitory or psychological effects that help in controlling the spread of COVID-19 pandemic in the society. Considering this, we formulate a *SEIQHR* mathematical model representing COVID-19 scenario with an incidence function of two infectious classes, namely symptomatic and asymptomatic with the inhibitory effect. The model is said to exhibit two equilibria, namely disease-free equilibrium (DFE) and endemic equilibrium (EE). Basic reproduction number is computed for the model. The local stability analysis is carried out for both the equilibria using Routh-Hurwitz criterion. The result shows stability of DFE when $R_0 < 1$ and persistence of COVID-19 when $R_0 > 1$. Sensitivity analysis of R_0 is also studied to understand the effect of various parameters used in modeling the spread of COVID-19. At the end, numerical results have been studied for the formulated model, showing existence of various bifurcations. The largest Lyapunov exponent is also calculated, indicating complexity of the model with low inhibitory rates.

Keywords: COVID-19, basic reproduction number, stability analysis, sensitivity analysis, bifurcation, maximum Lyapunov exponent.

13.1 Introduction

The first case of COVID-19, also known as corona virus disease 2019, originated in Wuhan, a small province of China in December 2019. After that, the virus continued to spread internationally and was not only confined to Wuhan. By now, corona virus has grabbed the whole world into his arms with 4,413,184 active cases and 533,847 deaths as on 5 July 2020 [12]. The human-to-human transmission of COVID-19 has made it a major concern for the world. In order to control the pandemic, it is important to understand the cause of spread and related symptoms. The virus spreads with close contact between humans mainly through the contact of small droplets that are released while coughing, sneezing, and talking. These droplets fall on the ground and when anyone comes in contact with these surfaces, the disease spreads. Some of the symptoms are fever, cough, fatigue, shortness of breath, and loss of smell [1, 3, 11, 12]. Further complication of COVID-19 disease leads to the acute respiratory distress syndrome [2].

To control the spread, some of the precautionary measures include: avoid contact with contaminated surfaces, make use of face mask, frequently wash hand, avoid touching nose, mouth, use sanitizers, etc. Also, governments worldwide have made use of various interventions like lockdown, curfew, ban on air travels, etc., to control the spread. This pandemic has also caused social and economic disruption globally.

To study the effectiveness of various interventions implemented by authorities and other related dynamics of COVID-19, huge amount of research work is carried out in terms of modeling of COVID-19. Mathematical modeling in epidemiology has a great impact in studying various dynamics of a disease. Some of the early mathematical models includes calculation of reproduction number by Cao [5] and formulation of SUQC (Susceptible, Un-quarantined, Quarantined and Confirmed infective) model by Zhao *et al.* [29]. Lin *et al.* [14] formulated a SEIR (Susceptible, Exposed, Infective and Recover) model considering individual and governmental action in controlling spread in China; also, Peng *et al.* [20] developed SEIR compartmental model to study epidemics of COVID-19 in China and Toda *et al.* [25] studied susceptible infective recover compartmental model.

In modeling epidemiology, to study the dynamics of the pandemic, various mathematical models with different types of incidence rate are studied. Incidence rate is defined as the rate at which susceptible becomes infectious. Some related epidemic models proposed include use of bilinear incidence rate βSI [8, 24, 27] and saturated incidence rate $\frac{\beta SI}{N}$ [13, 15, 17].

Capasso and Serio [6] were the first to introduce the saturated incidence rate $\frac{\beta SI}{1+\alpha I}$, with α being the inhibitory rate. It includes the behavioral change or crowding effect of infectives or preventive measures taken up by the susceptible leading to saturation stage of the epidemic, due to which unboundedness of contacts can be prevented, making saturated incidence rate reasonable than bilinear incidence rate. The other related research includes the following. Safi *et al.* [23] studied SEIR model with Holling type II incidence function. Ghosh *et al.* [9] studied logistic growth rate in susceptible for SIRS (Susceptible, Infective, Recover and Susceptible) model with inhibitory effect in infectives.

Followed by the above literature review and motivated by some of the works of [19, 22, 23], in this chapter, we formulate an *SEIQHR* compartmental model with inhibitory effect. Inhibitory factors play a role while susceptible transmits disease by contacting infectious individual (including both symptomatic and asymptomatic). We consider force of infection with inhibitory effect to get more realistic results related to COVID-19 pandemic throughout the world.

The chapter is organized as follows. Section 13.2 deals with the description and formulation of the mathematical model. In Section 13.3, basic reproduction number is calculated and the existence of unique endemic equilibria (EE) is established. The existing two equilibrium points are proved to be locally asymptotically stable under certain conditions established in Section 13.4. In Section 13.5, sensitivity analysis is carried out. In Section 13.6, numerical results are established and the chapter concludes in Section 13.7.

13.2 Description of Mathematical Model

Here we formulate *SEIQHR* mathematical model to study various dynamics of COVID-19 based on some realistic assumptions. The human population is divided into seven compartments as follows: S- susceptible class, E_{co}-class of individuals exposed to COVID-19, I_S- symptomatic, I_A- asymptomatic, Q- quarantine, H- hospitalized, and R- recovered or removed.

The notations and parametric values used for this COVID-19 model are given in Table 13.1.

The formulation of the model is as follows: here, we have considered new recruitments of individuals to susceptible class at a rate B and μ is taken as the natural death rate from all the compartments. Now susceptible population gets exposed to COVID-19 after coming in contact with infected individuals

Table 13.1 Parametric values with the description.

Notations	Description	Parametric Values	References
B	Birth rate	0.05	Assumed
β_1	Rate at which susceptible population gets exposed to COVID-19	0.72	[16]
β_2	Rate at which asymptomatic goes for quarantine	0.4	Assumed
β_3	Rate at which asymptomatic joins symptomatic class	0.38	Assumed
β_4	Rate at which symptomatic gets hospitalized	0.55	Assumed
β_5	Rate at which quarantine individual gets recovered	0.25	Assumed
β_6	Rate at which asymptomatic gets recovered	0.139	Calculated
β_7	Rate at which asymptomatic is admitted to the hospital	0.15	Assumed
β_8	Rate at which individuals in hospital gets recovered	0.5	[18]
θ_1, θ_2	Inhibitory coefficients	0.02, 0.04	Assumed
m	Rate at which exposed become infectious (both symptomatic and asymptomatic)	0.58	[18]
γ	Rate at which exposed become symptomatic	0.04	Assumed
μ	Natural morbidity rate	0.007	Assumed
μ_{co}	Morbidity rate due to COVID-19	0.3	[18]

either with symptomatic or asymptomatic. Here, spread of COVID-19 is taken as human-to-human transmission. The transmission rate of susceptible to exposed class occurs at a rate $\lambda(t)S$, where

$$\lambda(t) = \beta_1 \left(\frac{I_S}{1 + \theta_1 I_S} + \frac{I_A}{1 + \theta_2 I_A} \right).$$

Here, β_1 is the effective contact rate (contact capable of leading to infection) [22, 23] and θ_1, θ_2 are taken as the inhibitory effect or rate at which individuals take precautionary measures like use of face mask, social distancing, not rubbing nose and face using hands, and hygienic environment (use of soap, sanitizers, etc.) for the respective classes [16]. Next, the term γm and $(1 - \gamma)m$ are the rates at which individuals exposed to COVID-19 move to symptomatic and asymptomatic classes, respectively. The remaining $(1 - \gamma)(1 - m)$ are assumed to be quarantined. Also, since asymptomatic individuals do not show any symptoms of COVID-19, indicating that they may show negative results on testing, following this, the government at some places like India is asking asymptomatic individuals to quarantine themselves and is not testing them. Therefore, this is assumed to occur at a rate β_2, and, as a result, very few asymptomatic individuals are hospitalized at a rate β_7 (which occurred in the beginning of the spread). Recovery of asymptomatic is assumed to occur at a rate β_6. Similarly, quarantine individuals are assumed to recover with a rate β_5. Symptomatic individuals get hospitalized at a rate β_4. No direct recovery of symptomatic individuals is considered as they have to be hospitalized. Recovery rate of hospitalized class is considered to occur at a rate β_8. Also, COVID-19 related death is assumed to occur in hospitalized class at a rate μ_{co}. Also, it is observed that there is a very less chance of an individual getting infected again once recovered from COVID-19. So, we do not consider further susceptibility to COVID-19 of recovered individuals.

The dynamical system representing Figure 13.1 is given by the following set of non-linear differential equations:

$$\begin{aligned}
\frac{dS}{dt} &= B - (\lambda + \mu)S, \\
\frac{dE_{co}}{dt} &= \lambda S - (\gamma m + (1-\gamma)m + (1-\gamma)(1-m) + \mu)E_{co}, \\
\frac{dI_S}{dt} &= \gamma m E_{co} + \beta_3 I_A - (\beta_4 + \mu)I_S, \\
\frac{dI_A}{dt} &= (1-\gamma)mE_{co} - (\beta_2 + \beta_3 + \beta_6 + \beta_7 + \mu)I_A, \\
\frac{dQ}{dt} &= (1-\gamma)(1-m)E_{co} + \beta_2 I_A - (\beta_5 + \mu)Q, \\
\frac{dH}{dt} &= \beta_4 I_S + \beta_7 I_A - (\beta_8 + \mu + \mu_{co})H, \\
\frac{dR}{dt} &= \beta_5 Q + \beta_6 I_A + \beta_8 H - \mu R,
\end{aligned} \tag{13.1}$$

where $\lambda = \beta_1\left(\frac{I_S}{1+\theta_1 I_S} + \frac{I_A}{1+\theta_2 I_A}\right)$ and initial conditions as $S(0) \geq 0$, $E_{co}(0) \geq 0,\ I_S(0) \geq 0,\ I_A(0) \geq 0,\ Q(0) \geq 0,\ H(0) \geq 0,\ R(0) \geq 0.$

The feasible region for the system (13.1) is given by

$$\Omega = \left\{(S, E_{co}, I_S, I_A, Q, H, R) \in \mathbb{R}^7_+ : \right.$$
$$\left. 0 \leq (S + E_{co} + I_S + I_A + Q + H + R) \leq \frac{B}{\mu}\right\}.$$

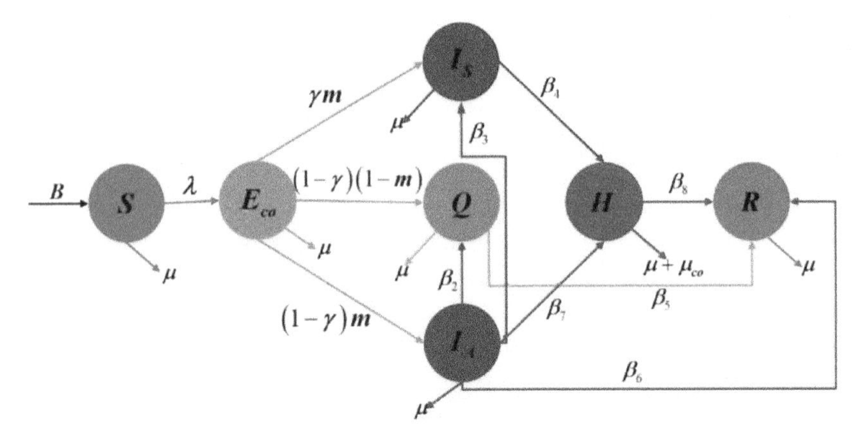

Figure 13.1 Flow of human population through compartments.

System (13.1) has the following equilibria:

1. Disease-free equilibrium (DFE) point $E^0\left(\frac{B}{\mu}, 0, 0, 0, 0, 0, 0\right)$, indicating absence of COVID-19 in the society.
2. EE point $E^*(S^*, E^*_{co}, I^*_S, I^*_A, Q^*, H^*, R^*)$, indicating existence of COVID-19, i.e., current situation of the world.

The existence of these two-equilibrium points is shown in the next section.

13.3 Basic Reproduction Number

The basic reproduction number is a dimensionless number measuring the spread of disease which plays a significant role in understanding and controlling the spread. It is defined as the number of secondary infections produced due to a single infected individual during its infectious period in a completely uninfected population. With the help of next generation matrix method described by Van den Driessche [26], we calculate basic reproduction number. We consider the exposed class and both infectious class for calculation as they directly affect the spread of COVID-19. The Jacobian matrices F and V are of new infections, and new transfer of individuals across the compartments respectively is calculated below.

$$F(E^0) = \begin{bmatrix} 0 & \frac{B\beta_1}{\mu} & \frac{B\beta_1}{\mu} \\ 0 & 0 & 0 \\ 0 & 0 & 0 \end{bmatrix},$$

$$V(E^0) = \begin{bmatrix} (\gamma m + (1-\gamma)m + (1-\gamma)(1-m) + \mu) & 0 & 0 \\ -\gamma m & \beta_4 + \mu & -\beta_3 \\ -(1-\gamma)m & 0 & \beta_2 + \beta_3 + \beta_6 + \beta_7 + \mu \end{bmatrix}.$$

The spectral radius of FV^{-1} at DFE point gives us the basic reproduction number R_0 as

$$R_0 = \frac{B\beta_1 m((\beta_3 + \beta_4 + \mu)(1-\gamma) + \gamma(\beta_2 + \beta_3 + \beta_6 + \beta_7 + \mu))}{\mu(\gamma m + (1-\gamma)m + (1-\gamma)(1-m))(\beta_2 + \beta_3 + \beta_6 + \beta_7 + \mu)(\beta_4 + \mu)}.$$

The R_0 for our model is calculated as 4.736 (using data of Table 13.1). Here, $R_0>1$ indicates that the disease will move on from one another, giving rise to the endemic situation of the spread of COVID-19, and, thus, the system (13.1) undergoes transcritical bifurcation and exchange of stability at $R_0 = 1$ [4, 10] as shown in Figure 13.3.

13.3.1 Existence of Unique Endemic Equilibrium Point

To find the conditions for existence of unique EE point E^* $(S^*,\ E_{co}^*,\ I_S^*,\ I_A^*,\ Q^*,\ H^*,\ R^*)$, the calculation is as follows:

At endemic point, we have

$$\lambda^*(t) = \beta_1 \left(\frac{I_S^*}{1+\theta_1 I_S^*} + \frac{I_A^*}{1+\theta_2 I_A^*} \right). \tag{13.2}$$

By setting the right-hand side of system (13.1) to zero, we get

$$S^* = \frac{B}{(\lambda^* + \mu)}, \quad E_{co}^* = \frac{\lambda^* B}{(\lambda^* + \mu)k_1}, \quad I_S^* = \frac{\lambda^* B(\gamma m k_3 + \beta_3(1-\gamma)m)}{(\lambda^* + \mu)k_1 k_2 k_3},$$

$$I_A^* = \frac{\lambda^* B(1-\gamma)m}{(\lambda^* + \mu)k_1 k_3}, \quad Q^* = \frac{\lambda^* B(1-\gamma)((1-m)k_3 + m\beta_2)}{(\lambda^* + \mu)k_1 k_3(\beta_5 + \mu)},$$

$$H^* = \frac{\lambda^* Bm(\beta_4(\gamma k_3 + \beta_3(1-\gamma)) + \beta_7 k_2(1-\gamma))}{(\lambda^* + \mu)k_1 k_2 k_3(\beta_8 + \mu + \mu_{co})},$$

$$R^* = \frac{\begin{array}{l}\lambda^* B((1-\gamma)k_2(\beta_8+\mu+\mu_{co})(\beta_5((1-m)k_3+m\beta_2) \\ +\beta_6 m(\beta_5+\mu))+m\beta_8(\beta_4(\gamma k_3+\beta_3(1-\gamma))+\beta_7 k_2(1-\gamma)) \\ (\beta_5+\mu))\end{array}}{(\lambda^*+\mu)k_1k_2k_3(\beta_8+\mu+\mu_{co})(\beta_5+\mu)\mu}.$$

$$k_1 = (\gamma m + (1-\gamma)m + (1-\gamma)(1-m) + \mu), \; k_2 = \beta_4 + \mu,$$
$$k_3 = \beta_2 + \beta_3 + \beta_6 + \beta_7 + \mu.$$

Now substituting the values of I_S^* and I_A^* in Equation (13.2), we obtain

$$\lambda^*(M_0(\lambda^*)^2 + M_1\lambda^* + M_2) = 0.$$

Here, for $\lambda^* = 0$, we get DFE point as mentioned in Section 13.2. For EE point, we have

$$M_0(\lambda^*)^2 + M_1\lambda^* + M_2 = 0, \tag{13.3}$$

with

$$M_0 = (Bm\theta_2(\beta_3(1-\gamma)+k_3\gamma)+k_1k_2k_3)(Bm\theta_1(1-\gamma)+k_1k_3),$$
$$M_1 = k_1^2k_2k_3^2\mu(1-R_0) - B^2m^2\beta_1(\theta_1+\theta_2)(1-\gamma)(\beta_3(1-\gamma)+k_3\gamma)$$
$$+k_1k_3\mu(Bm(k_2\theta_1+\theta_2(\beta_3(1-\gamma)+k_3\gamma))+k_1k_2k_3),$$
$$M_2 = k_1^2k_2k_3^2\mu^2(1-R_0).$$

Applying Descartes' rule of signs [28] to the quadratic Equation (13.3), there is a positive solution for λ^* if any of the following holds:

1. $M_0 > 0, \; M_1 > 0,$ and $M_2 < 0.$
2. $M_0 > 0, \; M_1 < 0,$ and $M_2 < 0.$

Hence, there exists a unique EE point for the system (13.1) if $R_0 > 1$.

13.4 Stability Analysis

In this section, we establish local stability for the equilibrium points of system (13.1).

Theorem 13.1. *The DFE point E^0 is said to be locally asymptotically stable if $R_0 < 1$ and it is unstable if $R_0 > 1$.*

Proof: The Jacobian of the system (13.1) at E^0 is given by

$$J^0 = \begin{bmatrix} -\mu & 0 & -\frac{B\beta_1}{\mu} & -\frac{B\beta_1}{\mu} & 0 & 0 & 0 \\ 0 & -(a_1+\mu) & \frac{B\beta_1}{\mu} & \frac{B\beta_1}{\mu} & 0 & 0 & 0 \\ 0 & \gamma m & -(\beta_4+\mu) & 0 & 0 & 0 & 0 \\ 0 & (1-\gamma)m & 0 & -(a_2+\mu) & 0 & 0 & 0 \\ 0 & (1-\gamma)(1-m) & 0 & \beta_2 & -(\beta_5+\mu) & 0 & 0 \\ 0 & 0 & \beta_4 & \beta_7 & 0 & -(\beta_8+\mu+\mu_{co}) & 0 \\ 0 & 0 & 0 & \beta_6 & \beta_5 & \beta_8 & -\mu \end{bmatrix},$$

where $a_1 = (\gamma m + (1-\gamma)m + (1-\gamma)(1-m))$, $a_2 = \beta_2 + \beta_3 + \beta_6 + \beta_7$.

The characteristic equation of J^0 is given by

$$(\Theta+\mu)^2(\Theta+\beta_5+\mu)(\Theta+\beta_8+\mu+\mu_{co})(\Theta^3+p_2\Theta^2+p_1\Theta+p_0) = 0$$

where

$$\begin{aligned} p_2 &= (a_1+\mu)+(a_2+\mu)+(\beta_4+\mu), \\ p_1 &= \left((a_1+\mu)(\beta_4+\mu)+(a_2+\mu)(\beta_4+\mu)+(a_1+\mu)(a_2+\mu)-\tfrac{B\beta_1 m}{\mu}\right), \\ p_0 &= (1-R_0)(a_1+\mu)(a_2+\mu)(\beta_4+\mu). \end{aligned}$$

The DFE is locally asymptotically stable if all its eigenvalues are negative or have negative real part. Clearly, if it satisfies Routh-Hurwitz stability criterion [21], i.e., $p_0 > 0, p_2 > 0$, and $p_2p_1 > p_0$ also $R_0 < 1$, it is stable.

Theorem 13.2. *The EE point E^* is asymptotically locally stable for*

$$\frac{S^*\beta_1\mu m\begin{pmatrix}(1-\gamma)\left(\frac{1}{1+\theta_1 I_A^*}-\frac{\theta_1 I_A^*}{(1+\theta_1 I_A^*)^2}\right)(\beta_4+\mu)\\ +\left(\frac{1}{1+\theta_1 I_S^*}-\frac{\theta_1 I_S^*}{(1+\theta_1 I_S^*)^2}\right)\\ (\beta_3+(\beta_2+\beta_6+\beta_7+\mu)\gamma)\end{pmatrix}}{\begin{array}{l}(\beta_4+\mu)(\gamma m+(1-\gamma)m+(1-\gamma)(1-m)+\mu)\\ (\beta_2+\beta_3+\beta_6+\beta_7+\mu)\left(\beta_1\left(\frac{I_S^*}{1+\theta_1 I_S^*}+\frac{I_A^*}{1+\theta_2 I_A^*}\right)+\mu\right)\end{array}}<1.$$

Proof: The Jacobian at EE is given by

$$J^*=\begin{bmatrix}-a_{11}-\mu & 0 & -a_{13} & -a_{14} & 0 & 0 & 0\\ a_{11} & -a_{22} & a_{13} & a_{14} & 0 & 0 & 0\\ 0 & \gamma m & -a_{33} & \beta_3 & 0 & 0 & 0\\ 0 & a_{42} & 0 & -a_{44} & 0 & 0 & 0\\ 0 & a_{52} & 0 & \beta_2 & -a_{55} & 0 & 0\\ 0 & 0 & \beta_4 & \beta_7 & 0 & -a_{66} & 0\\ 0 & 0 & 0 & \beta_6 & \beta_5 & \beta_8 & -\mu\end{bmatrix},$$

where

$$a_{11}=\beta_1\left(\frac{I_S^*}{1+\theta_1 I_S^*}+\frac{I_A^*}{1+\theta_2 I_A^*}\right),$$

$$a_{13}=S^*\beta_1\left(\frac{1}{1+\theta_1 I_S^*}-\frac{\theta_1 I_S^*}{(1+\theta_1 I_S^*)^2}\right),$$

$$a_{14}=S^*\beta_1\left(\frac{1}{1+\theta_1 I_A^*}-\frac{\theta_1 I_A^*}{(1+\theta_1 I_A^*)^2}\right),$$

$$a_{22}=(\gamma m+(1-\gamma)m+(1-\gamma)(1-m)+\mu),\quad a_{33}=\beta_4+\mu,$$

$$a_{44}=\beta_2+\beta_3+\beta_6+\beta_7+\mu,\quad a_{55}=\beta_5+\mu,$$

$$a_{42}=m(1-\gamma),a_{52}=(1-\gamma)(1-m),\quad a_{66}=\beta_8+\mu+\mu_{co}.$$

$$tr(J^*)=-(a_{11}+2\mu+a_{22}+a_{33}+a_{44}+a_{55}+a_{66})<0,$$

and principle minors of J^* are

$$M_1=-(a_{11}+\mu)<0,$$
$$M_2=a_{22}(a_{11}+\mu)>0,$$

$$M_3 = -(a_{22}a_{33}(a_{11} + \mu) - a_{13}\mu\gamma m) < 0,$$
$$M_4 = (a_{22}a_{33}a_{44}(a_{11} + \mu) - \mu(a_{42}(a_{14}a_{33} + a_{13}\beta_3) + a_{44}a_{13}\gamma m)) = t_1 > 0,$$
$$M_5 = -a_{55}t_1 < 0,$$
$$M_6 = a_{55}a_{66}t_1 > 0,$$
$$M_7 = -a_{55}a_{66}\mu t_1 < 0.$$

All the minors have alternative positive negative sign for $t_1 > 0$. Clearly, all the eigenvalues have negative real part or are negative if $(a_{22}a_{33}a_{44}(a_{11} + \mu) > \mu(a_{42}(a_{14}a_{33} + a_{13}\beta_3) + a_{44}a_{13}\gamma m))$. Hence, the EE point is locally asymptotically stable.

13.5 Sensitivity Analysis

In this section, we calculate sensitivity indices of R_0 related to only those parameters which are used in formulation of R_0 using the normalized formula $\gamma_\alpha^{R_0} = \frac{\partial R_0}{\partial \alpha} \cdot \frac{\alpha}{R_0}$, where α is the model parameter [7]. The study of sensitivity in epidemiology gives the distribution of different parameters that affect the spread of a disease. The description and values calculated are tabulated in Table 13.2.

Table 13.2 Analysis of parameters in the spread of COVID-19.

Parameter	Value	Description
B	1	The spread of COVID-19 is directly proportional to birth rate
β_1	1	This indicates that if β_1 increases by 10%, then R_0also increases by 10%
β_2	–0.3829	By quarantining of asymptomatic individuals, R_0can be reduced by 38%
B	1	The spread of COVID-19 is directly proportional to birth rate
β_3	0.0466	With the increases in symptomatic individuals from asymptomatic class, it increases R_0 by 4.66%
β_4	–0.3755	Hospitalizing symptomatic individuals decrease R_0 by 37.5%
β_6	–0.1331	Recovery of asymptomatic individuals decreases R_0 by 13%
β_7	–0.1436	Hospitalizing asymptomatic individuals decreases R_0 by 14%
γ	0.0234	Symptomatic individuals increase R_0 by 2.34
m	0.9761	The rate at which individuals become infectious causes a larger increase in R_0
μ	–1.0190	Natural mortality rate highly affects R_0

From Table 13.2, we observe that m-rate of infectiousness has the most positive impact on the spread of COVID-19, whereas μ-natural death rate has the most negative impact on the spread of COVID-19.

13.6 Numerical Simulation

In this section, we perform simulation for the system (13.1) using numerical data given in Table 13.1. The data has been prepared and assumed in the sense that it represents the COVID-19 scenario by referring previous research work [12, 16, 18]. All the simulation has been carried out using MATLAB software. For simulation purpose, we have considered $\mu = 0.07$ and the remaining parameters are kept the same as in Table 13.1. We have computed trajectories showing stability around endemic point, various bifurcation diagrams, periodic solution, and chaos.

In Figure 13.2, we plot trajectories of various compartments considered in the model. Here, we observe oscillations in the various compartments. We observe that the symptomatic individuals are hospitalized at a larger scale than the asymptomatic population, indicating that COVID-19 continues to spread even after various measures taken up by the government due to non-hospitalization of asymptomatic individuals. Further increase in hospitalization of symptomatic individuals leads to chaos as shown in Figure 13.4. Also, Figure 13.2 shows the stability around coexistence of EE point E^* with time.

In Figure 13.3, we plot transcritical bifurcation diagram for $R_0 > 1$, showing the existence of bifurcation and exchange of stability occurring at $R_0 = 1$. At $R_0 < 1$, we have $I_A^* = 0$ (class of asymptomatic individuals), and, thus, DFE is stable, and, at $R_0 > 1$, DFE is unstable. At $R_0 > 1$, we have $I_A^* > 0$ for which EE is stable. As a result, COVID-19 persists in society.

In Figure 13.4, we observe, by increasing hospitalization of symptomatic individuals, that the system depicts periodic solution, indicating chaos. Therefore, this scenario relates to real life, showing existence of chaos in hospitals, such as for beds, ventilators, etc., if there is increase in hospitalization to a greater extend.

In Figure 13.5, we plot Lyapunov exponents for the system (13.1) with time, where $B = 50\,, \beta_1 = 0.8, m = 0.67, \beta_4 = 1.55$, and the remaining values are kept the same as in Table 13.1. Here, we use Wolf's Algorithm to compute Lyapunov exponents. The maximum Lyapunov exponent being positive indicates the chaotic nature of dynamical system.

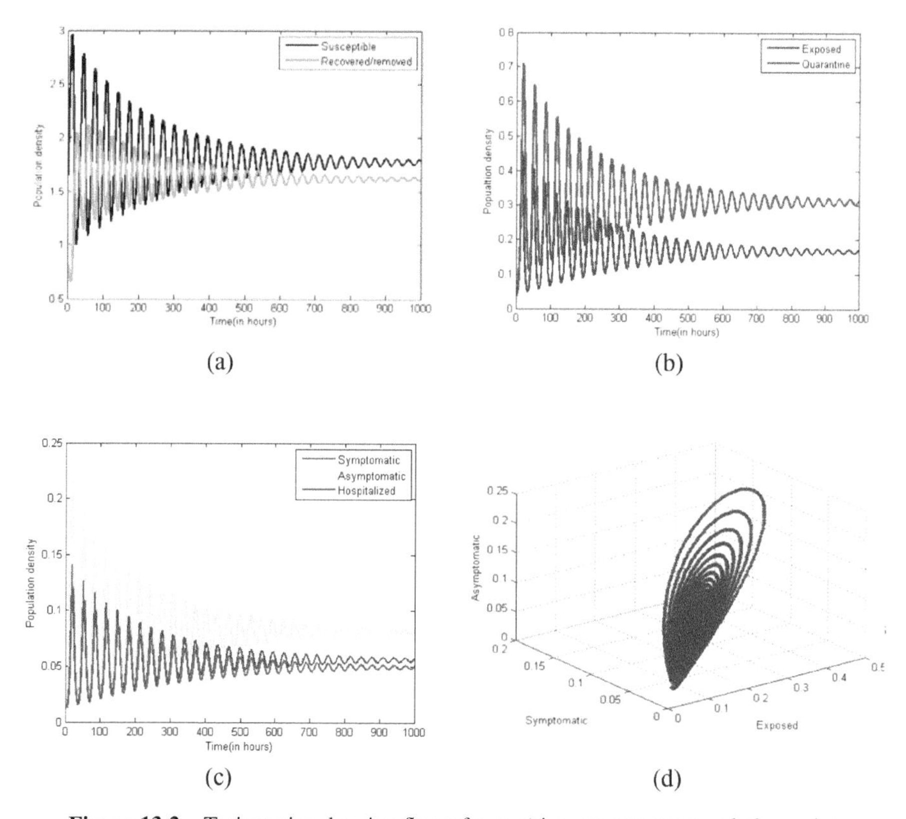

Figure 13.2 Trajectories showing flow of respective compartment and phase plot.

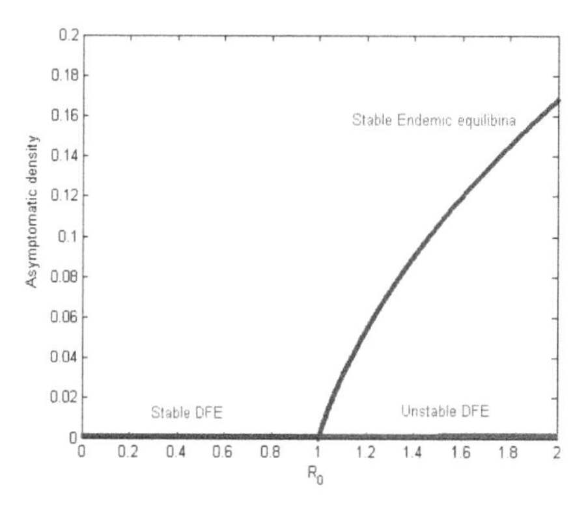

Figure 13.3 Bifurcation showing exchange of stability at $R_0 = 1$ with $\theta_1 = 10.02$, $\theta_2 = 5.04$, and other parameters fixed as in Table 13.1.

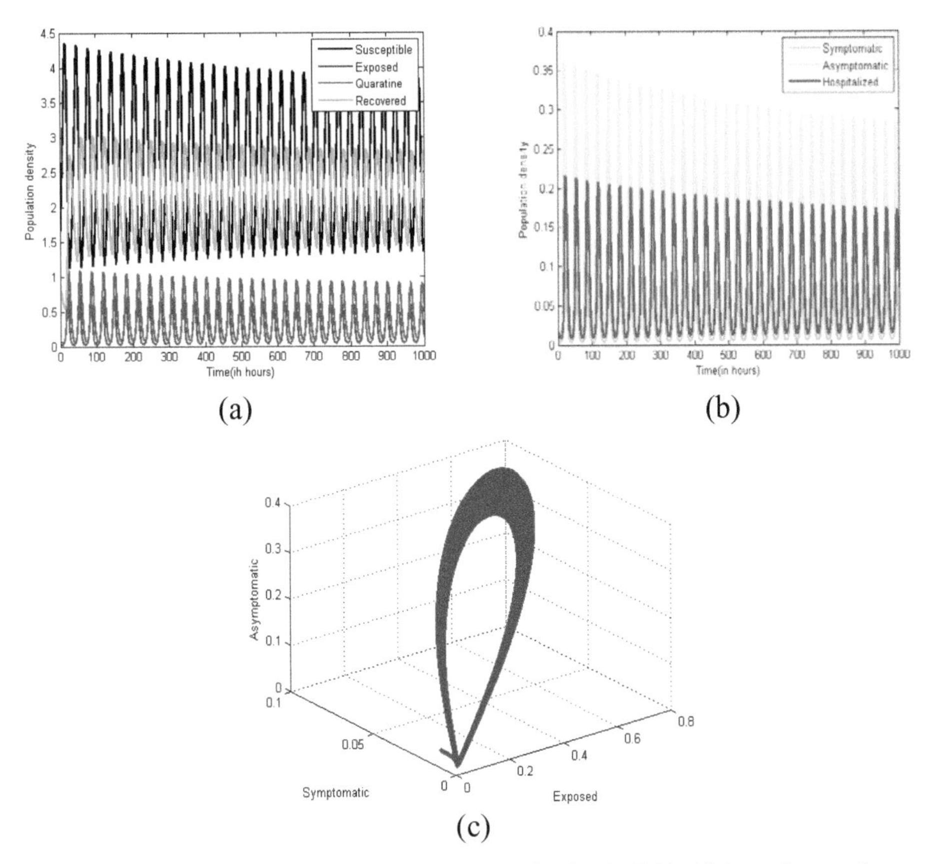

Figure 13.4 For $\beta_4 = 1.55$ and other parameters fixed as in Table 13.1,we observe chaos around endemic equilibrium point.

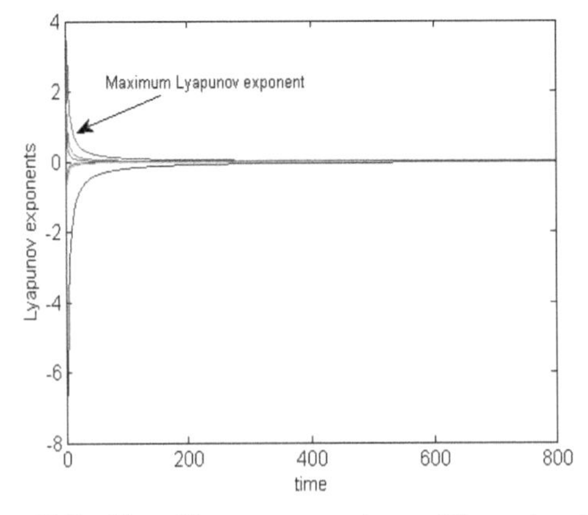

Figure 13.5 Plot of Lyapunov spectrum of the system (13.1).

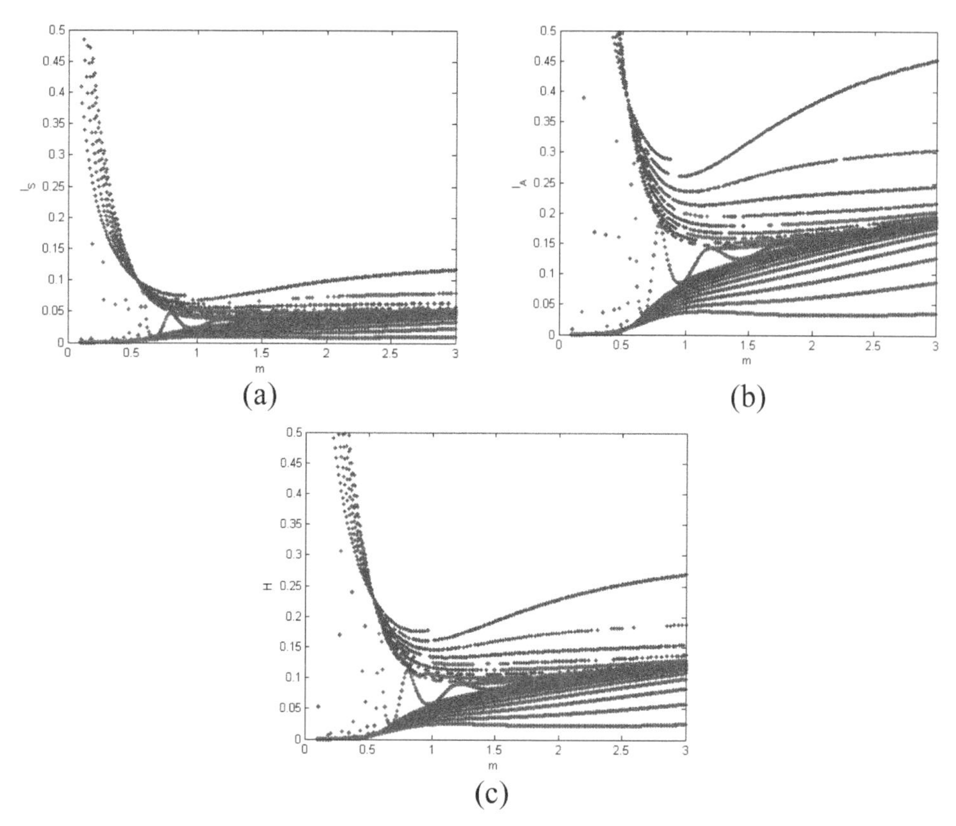

Figure 13.6 Bifurcation diagram for symptomatic, asymptomatic, and hospitalized population with the bifurcation parameter m for $\beta_4 = 1.55$.

In Figure 13.6, we observe that with the increase in m, i.e., rate of infection, the system turns chaotic. Also, in the interval $[0, 1]$, we have periodic solution around an EE.

In Figure 13.7, we plot bifurcation diagrams for symptomatic (I_S), asymptomatic (I_A), and hospitalized population (H) with inhibitory parameter. Here, blue and red lines indicate the maximum and minimum values of oscillations. We observe that after a threshold value at 0.4 (approximately), with the further increase in θ, the respective classes get stabilized.

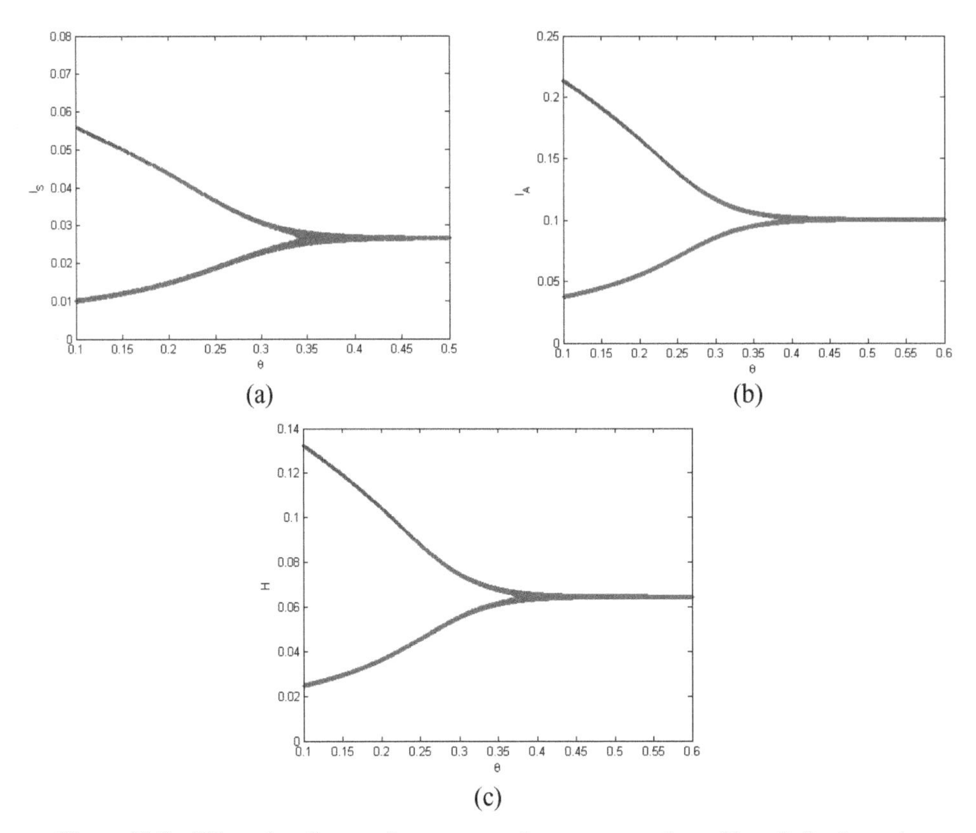

Figure 13.7 Bifurcation diagram for symptomatic, asymptomatic, and hospitalized population with the parameter $\theta = \theta_1 = \theta_2 = 0.04$.

13.7 Conclusion

In this chapter, a non-linear *SEIQHR* model is formulated with two infectious classes, namely symptomatic and asymptomatic. Here, we considered the transmission of infection among susceptible individuals through both the infectious classes. Now to get the realistic picture of the model, we adopt two rates defined as inhibitory factor, or say precautionary measures, into the model within the incidence function $\lambda(t)$. For the formulated model, we have computed two equilibrium points, namely DFE and EE points. The basic reproduction number is calculated as $R_0 = 4.763 > 1$, indicating persistence of COVID-19, following which the system undergoes transcritical bifurcation at $R_0 = 1$. Also, local stability is worked out using Routh-Hurwitz criterion for both the equilibria. Next, we have also analyzed sensitivity of R_0 with different model parameters, which helps us in providing controlling procedures for the epidemic. In simulation section, we have observed, through transcritical bifurcation, the existence of stable DFE for $R_0 < 1$ and stable EE point for $R_0 > 1$. Oscillations have also been obtained for small values of inhibitory factors. Also, the system gets periodic solutions and chaos with the increase in hospitalization of symptomatic individuals and for small values of inhibitory factor. The chaotic nature of system is also verified by calculating maximum Lyapunov exponent. Also, it is concluded that with the increase in inhibitory factors, the system gets stable observed through bifurcation in Figure 13.7. Now, since there is no vaccine available, it is important to adopt precautionary measures as mentioned inhibitory parameters since those are the ones which can make the system free of chaos; i.e., one should always wash hands frequently, always wear mask while steeping out of home, avoid touching nose and mouth frequently, etc. As of 5 July 2020, since lockdown has ended in various countries, it is now each individual's responsibility to take care of themselves.

Acknowledgement

The authors thank DST-FIST file # MSI-097 for technical support to the department. N. Sheoran would like to extend sincere thanks to the Education Department, Gujarat State for providing scholarship under ScHeme Of Developing High quality research (SHODH). E. Jayswal is funded by UGC granted National Fellowship for Other Backward Classes (NFO-2018-19-OBC-GUJ-71790).

13.8 Conflict of Interest

The authors do not have any conflict of interest.

References

[1] "Coronavirus Disease 2019 (COVID-19)—Symptoms". U.S. Centers for Disease Control and Prevention (CDC).

[2] "Interim Clinical Guidance for Management of Patients with Confirmed Coronavirus Disease (COVID-19)". U.S. Centers for Disease Control and Prevention (CDC).

[3] "Q & A on coronaviruses (COVID-19)". World Health Organization.

[4] Buonomo B, 'A note on the direction of the transcritical bifurcation in epidemic models', Nonlinear Anal. Model. Control., pp. 38–55, Italy, Jan., 2015.
[5] Z. Cao, Q. Zhang, X. Lu, D. Pfeiffer, Z. Jia, H. Song, D. D. Zeng, 'Estimating the effective reproduction number of the 2019-nCoV in China', China, Jan., 2020.
[6] V. Capasso, G. Serio, 'A generalization of the Kermack-McKendrick deterministic epidemic model', Math. Biosci., pp. 43–61, Italy, Nov., 1978.
[7] N. Chitnis, J. M. Hyman, J. M. Cushing, 'Determining important parameters in the spread of malaria through the sensitivity analysis of a mathematical model', Bull. Math. Biol., pp. 1272–1296, July, 2008.
[8] Y. Enatsu, E. Messina, Y. Muroya, Y Nakata, E. Russo, A. Vecchio, 'Stability analysis of delayed SIR epidemic models with a class of nonlinear incidence rates', Appl. Math. Comput., pp. 5327–5336, USA, Jan., 2012.
[9] J. K. Ghosh, U. Ghosh, S. Sarkar, S. 'Qualitative Analysis of Both Hyperbolic and Non-hyperbolic Equilibria of a SIRS Model with Logistic Growth Rate of Susceptibles and Inhibitory Effect in the Infection', CMST, pp. 285–300, India, 2018.
[10] J. Guckenheimer and P. Holmes, 'Nonlinear oscillations, Dynamical systems, and Bifurcations of vector fields', Springer Science and Business Media, USA, Nov., 2013.
[11] C. Hopkins, 'Loss of sense of smell as marker of COVID-19 infection'. Ear, Nose and Throat surgery body of United Kingdom'.
[12] https://www.worldometers.info/coronavirus/ (accessed on 5th July 2020, 15:59 PM IST)
[13] Z. Hu, S. Liu, H. Wang, 'Backward bifurcation of an epidemic model with standard incidence rate and treatment rate', Nonlinear Anal. Real World Appl., pp. 2302–2312, Dec., 2008.
[14] Q. Lin, S. Zhao, D. Gao, Y. Lou, S. Yang, S. Musa, S.S., ... and D. He, 'A conceptual model for the outbreak of Coronavirus disease 2019 (COVID-19) in Wuhan, China with individual reaction and governmental action', Int. J. Infect. Dis., pp. 211–216, China, Mar., 2020.
[15] J. Liu, 'Bifurcation analysis for a delayed SEIR epidemic model with saturated incidence and saturated treatment function', J. Biol. Dynam., pp. 461–480, China, Jan., 2019.

[16] M. Mandal, S. Jana, S.K. Nandi, A. Khatua, S. Adak, T.K. Kar, 'A model-based study on the dynamics of COVID-19: Prediction and control', Chaos Solitons Fractals, pp. 1–12, India, May, 2020.
[17] Y. Nakata and T. Kuniya, 'Global dynamics of a class of SEIRS epidemic models in a periodic environment', J. Math. Anal. Appl., 230–237, Japan, March, 2010.
[18] F. Ndairou, I. Area, J.J. Nieto, D.F. Torres, 'Mathematical modelling of covid-19 transmission dynamics with a case study of Wuhan', Chaos Solitons Fractals, pp. 1–6, Spain, Apr., 2020.
[19] D.A. Oluyori, H.O. Adebayo, 'Global Analysis of an SEIRS Model for COVID-19 Capturing Saturated Incidence with Treatment Response', medRxiv, Nigeria, Jan., 2020.
[20] L. Peng, W. Yang, D. Zhang, C. Zhuge and L. Hong, 'Epidemic analysis of COVID-19 in China by dynamical modeling', arXiv preprint arXiv:2002.06563, Feb., 2020.
[21] E.J. Routh, 'A treatise on the stability of a given state of motion: particularly steady motion', Macmillan and Company, 1877.
[22] M.A. Safi, 'Global Stability Analysis of Two-Stage Quarantine-Isolation Model with Holling Type II Incidence Function', Mathematics, pp. 1–12, Jordan, Apr., 2019.
[23] M.A. Safi, S.M. Garba, 'Global stability analysis of SEIR model with holling type II incidence function', Comput. Math. Methods Med., pp. 1–12, South Africa, Sept., 2012.
[24] Y. Takeuchi, W. Ma, E. Beretta, 'Global asymptotic properties of a delay SIR epidemic model with finite incubation times', Nonlinear Anal. Theory Methods Appl., pp. 931–94, Japan, Nov., 2000.
[25] A.A. Toda, 'Susceptible-infected-recovered (SIR) dynamics of Covid-19 and economic impact1. Covid Economics', arXiv preprint arXiv:2003.11221, Mar., 2020.
[26] P. Van den Driessche, J. Watmough, 'Reproduction numbers and sub-threshold endemic equilibria for compartmental models of disease transmission', Math. biosci., pp. 29–48, Canada, Nov., 2000.
[27] J.J. Wang, J.Z. Zhang, Z. Jin, 'Analysis of an SIR model with bilinear incidence rate', Nonlinear Anal. Real World Appl., pp. 2390–2402, China, Aug., 2010.
[28] X. Wang, 'A simple proof of Descartes's rule of signs', Am. Math. Mon., pp. 525, New York, June, 2004.
[29] S. Zhao and H. Chen, 'Modeling the epidemic dynamics and control of COVID-19 outbreak in China', Quant Biol., pp. 1–9, China, Mar., 2020.

Index

About the Editors

Akshay Kumar received the B.Sc. and M.Sc. degrees in science from Chaudhary Charan Singh University, Meerut, India, in 2010 and 2012, respectively, and the Ph.D. degree major in mathematics and minor in computer science from G. B. Pant University of Agriculture and Technology, Pantnagar, India, in 2017.

He is currently working as an Assistant Professor with the Department of Mathematics, Graphic Era Hill University, Dehradun, India. He is a regular Reviewer for international journals including Elsevier, Springer, Emerald, Cambridge University Press, World Scientific, IGI Global Publisher, and many other publishers. He has published 25 research papers in IEEE, Emerald, Inder science, *International Journal of Mathematical, Engineering and Management Sciences*, IGI Global Publisher, and many other national and international journals of repute and also presented his works at national and international conferences. His fields of research are in reliability theory and applied mathematics.

Mangey Ram received the Ph.D. degree major in mathematics and minor in computer science from G. B. Pant University of Agriculture and Technology, Pantnagar, India.

He has been a Faculty Member for around 12 years and has taught several core courses in pure and applied mathematics at undergraduate, postgraduate, and doctorate levels. He is currently the *Research Professor* with Graphic Era (Deemed to be University), Dehradun, India. Before joining the Graphic Era, he was a Deputy Manager (Probationary Officer) with Syndicate Bank for a short period. He is the Editor-in-Chief of *International Journal of Mathematical, Engineering and Management Sciences* and *Journal of Reliability and Statistical Studies*, and Editor-in-Chief of six Book Series with *Elsevier, CRC Press-A Taylor and Frances Group, WalterDe Gruyter Publisher Germany*, and *River Publisher* and the Guest Editor and Member of the editorial board of various journals. He has published more than 225 research publications (journal articles/books/book chapters/conference articles) in *IEEE,*

Taylor & Francis, *Springer*, *Elsevier*, *Emerald*, *World Scientific*, and many other national and international journals and conferences. Also, he has published more than 50 books (authored/edited) with international publishers like *Elsevier*, *Springer Nature*, *CRC Press-A Taylor and Frances Group*, *WalterDe Gruyter Publisher Germany*, and *River Publisher*. His fields of research are in reliability theory and applied mathematics. Dr. Ram is a Senior Member of the IEEE and a Senior Life Member of Operational Research Society of India, Society for Reliability Engineering, Quality and Operations Management in India, Indian Society of Industrial and Applied Mathematics. He has been a member of the organizing committee of a number of international and national conferences, seminars, and workshops. He has been conferred with "*Young Scientist Award*" by the Uttarakhand and State Council for Science and Technology, Dehradun, in 2009. He has been awarded the "*Best Faculty Award*" in 2011; "Research Excellence Award" in 2015; and, recently, "*Outstanding Researcher Award*" in 2018 for his significant contribution in academics and research at Graphic Era Deemed to be University, Dehradun, India.

Hari Mohan Srivastava was born on 05 July 1940, in Karon (District Ballia) in the Province of Uttar Pradesh in India. He began his university-level teaching career right after having received the M.Sc. degree in mathematics in the year 1959 at the age of 19 years. He earned the Ph.D. degree in 1965 while he was a full-time member of the teaching faculty at the Jai Narain Vyas University of Jodhpur in India (since 1963). Currently, Professor Srivastava holds the position of Professor Emeritus in the Department of Mathematics and Statistics at the University of Victoria in Canada, having joined the faculty there in 1969. Professor Srivastava has held (and continues to hold) numerous Visiting, Honorary, and Chair Professorships at many universities and research institutes in different parts of the world. Having received several D.Sc. (*honoris causa*) degrees as well as honorary memberships and fellowships of many scientific academies and scientific societies around the world, he is also actively associated editorially with numerous international scientific research journals as an Honorary or Advisory Editor or as an Editorial Board Member. He has also edited (and is currently editing) many Special Issues of scientific research journals as the Lead or Joint Guest Editor, including (for example) the MDPI journals, *Axioms*, *Mathematics*, and *Symmetry*, the Elsevier journals, *Journal of Computational and Applied Mathematics*, *Applied Mathematics and Computation*, *Chaos, Solitons & Fractals*, *Alexandria Engineering Journal*, and *Journal of King Saud*

University – Science, the Wiley journal, *Mathematical Methods in the Applied Sciences*, the Springer journals, *Advances in Difference Equations*, *Journal of Inequalities and Applications*, *Fixed Point Theory and Applications*, and *Boundary Value Problems*, the American Institute of Physics journal, *Chaos: An Interdisciplinary Journal of Nonlinear Science*, the American Institute of Mathematical Sciences journal, *AIMS Mathematics*, the Hindawi journals, *Advances in Mathematical Physics*, *International Journal of Mathematics and Mathematical Sciences*, and *Abstract and Applied Analysis*, the De Gruyter (*now* the Tbilisi Centre for Mathematical Sciences) journal, *Tbilisi Mathematical Journal*, the Yokohama Publisher journal, *Journal of Nonlinear and Convex Analysis*, the University of Nis journal, *Filomat*, the Ministry of Communications and High Technologies (Republic of Azerbaijan) journal, *Applied and Computational Mathematics: An International Journal*, and so on. He is a Clarivate Analytics [Thomson Reuters] (Web of Science) Highly-Cited Researcher.

Professor Srivastava's research interests include several areas of pure and applied mathematical sciences, such as (for example) real and complex analysis, fractional calculus and its applications, integral equations and transforms, higher transcendental functions and their applications, q-series and q-polynomials, analytic number theory, analytic and geometric inequalities, probability and statistics, and inventory modeling and optimization. He has published 36 books, monographs, and edited volumes, 36 book (and encyclopedia) chapters, 48 papers in international conference proceedings, and more than 1350 peer-reviewed international scientific research journal articles, as well as Forewords and Prefaces to many books and journals.

Further details about Professor Srivastava's professional achievements and scholarly accomplishments, as well as honors, awards, and distinctions can be found at the following Web Site:

http://www.math.uvic.ca/~harimsri/